System: Contemporary Concept, Definition, and Language

-- SBC Architecture at Work --

William S. Chao

2

CONTENTS

PREFACE

Human beings have employed the concept, definition, and language of systems so widely in all kinds of scientific studies. A systems definition is used to describe what a system is. Systems definition 1.0 defines a system hopefully to be an integrated whole, embodied in its components, their interrelationships with each other and the environment, and the principles and guidelines governing its design and evolution.

Systems structure and systems behavior are the two most significant views of a system. In order to achieve an integrated whole of a system, we first need to integrate the systems structure and behavior. Because systems definition 1.0 defining a system does not describe the integration of systems structure and systems behavior, very likely it will never be able to form a truly integrated whole of a system. In this situation, systems definition 1.0 is powerless in defining a system appropriately.

Structure-behavior coalescence (SBC) architecture provides an elegant way to integrate the structure and behavior of a system. Therefore, systems definition 2.0 shall use the SBC architecture to define a system. Systems definition 2.0 defines a system, through SBC architecture, truly to be an integrated whole, embodied in its components, their interactions with each other and the environment, and the principles and guidelines governing its design and evolution.

In this book, we give the contemporary concept and definition of a system based on the SBC architecture. By this book's introduction and elaboration of the SBC architecture, all readers shall understand clearly how the SBC architecture helps us effectively define a system.

ABOUT THE AUTHOR

Dr. William S. Chao is the CEO & founder of SBC Architecture International®. SBC (Structure-Behavior Coalescence) architecture is a systems architecture which demands the integration of systems structure and systems behavior of a system. SBC architecture applies to hardware architecture, software architecture, enterprise architecture, knowledge architecture, and thinking architecture. The core theme of SBC architecture is: "Architecture = Structure -->> Behavior."

William S. Chao received his bachelor degree (1976) in telecommunication engineering and master degree (1981) in information engineering, both from the National Chiao-Tung University, Taiwan. From 1976 till 1983, he worked as an engineer at Chung-Hwa Telecommunication Company, Taiwan.

William S. Chao received his master degree (1985) in information science and Ph.D. degree (1988) in information science, both from the University of Alabama at Birmingham, USA. From 1988 till 1991, he worked as a computer scientist at GE Research and Development Center, Schenectady, New York, USA.

Dr. William S. Chao has been teaching at National Sun Yat-Sen University, Taiwan since 1992 and now serves as the president of Association of Enterprise Architects, Taiwan Chapter. His research covers: systems architecture, hardware architecture, software architecture, enterprise architecture, knowledge architecture, and thinking architecture.

PART I: BASIC IDEAS

Chapter 1: Introduction to Systems

The word "system" originates from the Greek term, systēma, meaning "composition" or "whole." The notion of systems has been so widely used in all kinds of scientific studies such as systems analysis and design [Hoff10, Shel11], systems architecting [Maie09, Mull11], systems architecture [Burd10, Roza11], systems bible [Gall03, Kill09], systems biology [Klip09, Voit12], system dynamics [Forr61, Ogat03, Palm09], systems ecology [Jorg12, Odum94], systems engineering [Beam90, Kass07, Koss11], systems medicine [Pork78, Weil04, Weil00], systems modeling [Frie11], systems physiology [Raff11, Sher09], systems requirement [Bere09, Grad06], systems science [Warf06], systems theory [Bert69, Luhm12], systems thinking [Chec99, Ghar11, Mead08], systems view [Bert81, Lasz96].

In this chapter, we first introduce the systems definition 1.0. We then introduce the physical and virtual systems. A physical system exists in the physical, concrete, or real world. A virtual system exists in the virtual, abstract, or notional world. A system has a boundary. The system itself is inside the boundary and the environment is outside the boundary. After that, we then introduce the high order systems. A system evolves when it changes and the final section of this chapter will introduce the evolution of a system.

1-1 Systems Definition 1.0

All things that strike us as something independent are essentially parts of a system. We usually call the parts of a system its components. Components are sometimes labeled as parts, entities, objects, building blocks, and non-aggregated systems [Chao14a, Chao14b, Chao14c].

In the 1920s, Ludwig von Bertalanffy wrote: there exist models, principles, and laws that apply to generalized systems or their subclasses, irrespective of their particular kind, the nature of their constituent elements, and the relationships or "forces" between them [Bert69]. In this book, we refer what Ludwig von Bertalanffy proposed and developed as the systems definition 1.0.

The need for defining a system arises because any real-life system is inherently complicated. It is impossible to comprehend fully the intricate interrelationships of any system of the real world with its environment, or to describe all its components and each of its details. Systems definition is an "artifact" created by humans to describe what a system is [Kapo94].

Every system is something the whole. Systems emphasize the holistic vision. Systems definition 1.0 defines a system, in Figure 1-1, hopefully to be an integrated whole, embodied in its components, their interrelationships with each other and the environment, and the principles and guidelines governing its design and evolution [Chec99, Ghar11, Mead08].

A system, hopefully is an integrated whole,
embodied in its components,
their interrelationships with each other and the environment,
and the principles and guidelines governing its design and evolution.

Figure 1-1   Systems Definition 1.0 Defining a System

A system defined by the systems definition 1.0 has the following characteristics: 1) hopefully, it is an integrated whole; 2) it is embodied in its assembled components; 3) components are interrelated with each other and the environment; 4) it evolves; and 5) it uses structural decomposition [Chao12, Ghar11] rather than functional decomposition [Scho10].

Systems definition is used to describe what a system is. Without a systems definition, everybody has his own saying about a system and never be able to reach a consensus. For example, John Irving thinks the *Wardrobe_A* is embodied in its assembled components of *Drawer_1* and *Drawer_2*, their interrelationships with each other and the environment; Sandra Woods thinks the *Wardrobe_A* is embodied in its assembled components of *Drawer_1*, *Drawer_2*, *Drawer_3*, and *Drawer_4*. It is impossible for John Irving and Sandra Woods to work together on the *Wardrobe_A* if they can not reach a common definition. To solve the conflict between John Irving and Sandra Woods, here comes the systems definition 1.0 defining the *Wardrobe_A*, shown in Figure 1-2, hopefully to be an integrated whole embodied in its assembled components of *Drawer_1*, *Drawer_2*, and *Drawer_3*, their interrelationships with each other and the environment, and the principles and guidelines governing its design and evolution.

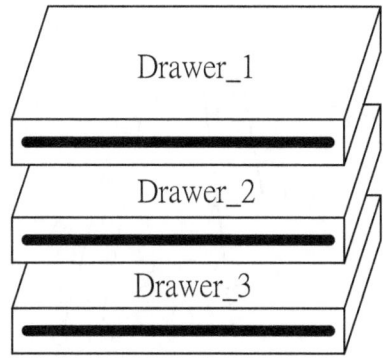

Figure 1-2   Systems Definition 1.0 Defining the *Wardrobe_A*

As a second example, systems definition 1.0 defines an *Eyeglasses*, shown in Figure 1-3, hopefully to be an integrated whole embodied in its assembled components of *Frames* and *Lenses*, their interrelationships with each other and the environment, and the principles and guidelines governing its design and evolution.

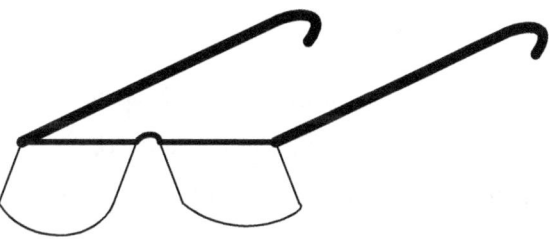

Figure 1-3   Systems Definition 1.0 Defining an *Eyeglasses*

As a third example, systems definition 1.0 defines a *Swing*, shown in Figure 1-4, to be hopefully an integrated whole embodied in its assembled components of *Ropes* and *Seat*, their interrelationships with each other and the environment, and the principles and guidelines governing its design and evolution.

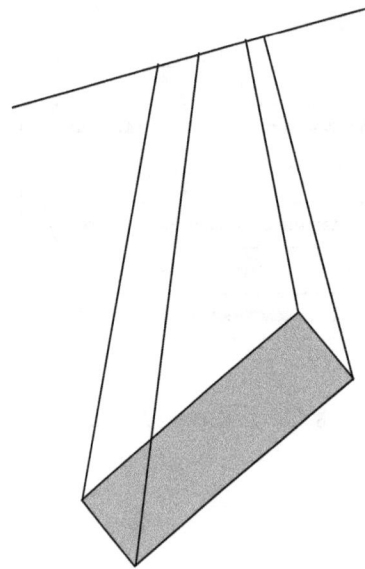

Figure 1-4   Systems Definition 1.0 Defining a *Swing*

1-2 Physical and Virtual Systems

In general, the system are divided into two categories: 1) physical systems and 2) virtual systems.

A physical system exists in the physical world [Acko68]. A physical system is also called a concrete or real system. For example, a *Bicycle* composed of *Wheels*, *Frame*, and *Pedal*, shown in Figure 1-5, is a physical, concrete, or real system.

Figure 1-5   A *Bicycle* is a Physical System

As a second example, a *Chair* composed of *Seat*, *Back*, and *Legs*, shown in Figure 1-6, is a physical, concrete, or real system.

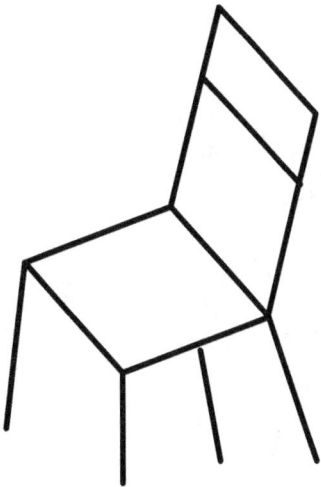

Figure 1-6   A *Chair* is a Physical System

A virtual system is a system that is composed of non-physical components, i.e., ideas, thoughts, or notions. A virtual system exists in the virtual, abstract, or notional world. For example, a fairy tale "*Jack and the Beanstalk*" composed of "*Jack*" and "*the Giant*," shown in Figure 1-7, is a virtual, abstract, or notional system.

Figure 1-7   *Jack and the Beanstalk* is a Virtual System

As a second example, For example, a software *Multi-Tier Personal Data System* composed of *MTPDS_GUI*, *Age_Logic*, *Overweight_Logic*, and *Personal_Database*, shown in Figure 1-8, is a virtual, abstract, or notional system.

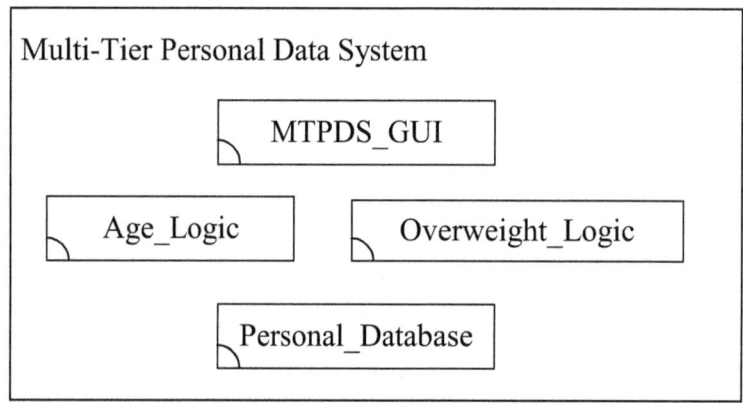

Figure 1-8   *Multi-Tier Personal Data System* is a Virtual System

1-3 Boundary and Environment of a System

We scope a system by defining its boundary as shown in Figure 1-9. All components of the system are inside the boundary while the environment is outside the boundary.

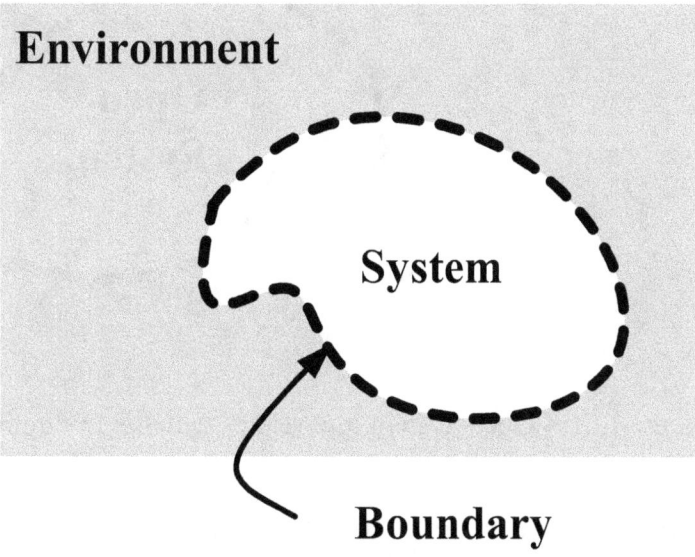

Figure 1-9   Boundary and Environment of a System

The environment is also known as the surroundings. A system may or may not interrelate with the environment. An open system shall interrelate with the environment through the exchange of matter, energy, data, information, or message as shown in Figure 1-10.

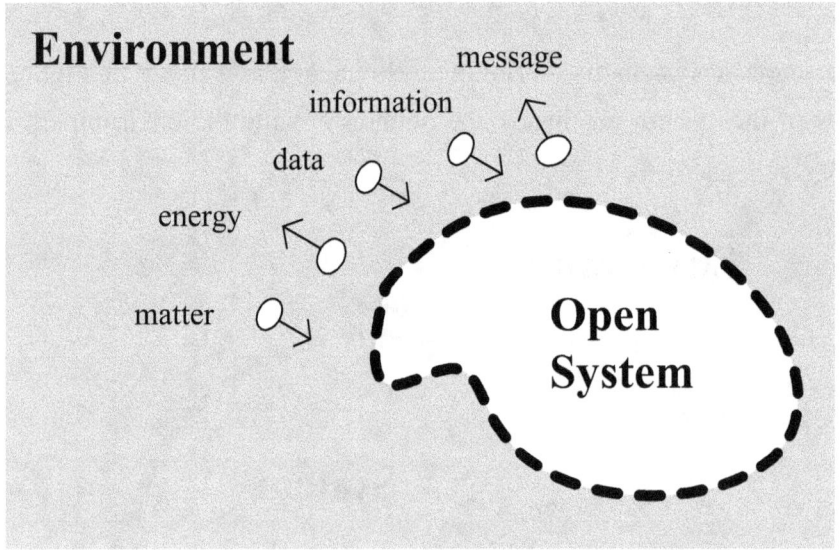

Figure 1-10   Open System Interrelates with the Environment

An isolated system does not interrelate with the environment at all. There is no exchange of matter, energy, data, information, or message between the isolated system and the environment as shown in Figure 1-11.

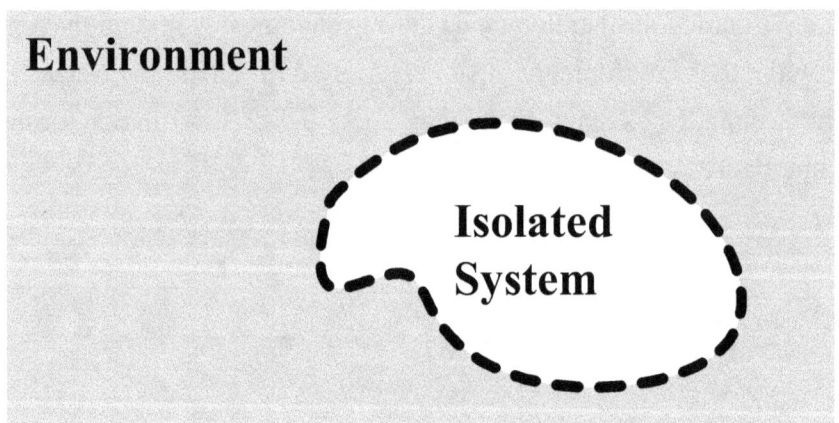

Figure 1-11   Isolated System Does Not Interrelate with the Environment

1-4 Higher-Order Systems

Higher-order systems interrelate with the environment through the exchange of not only matter, energy, data, information, or message but also systems as shown in Figure 1-12.

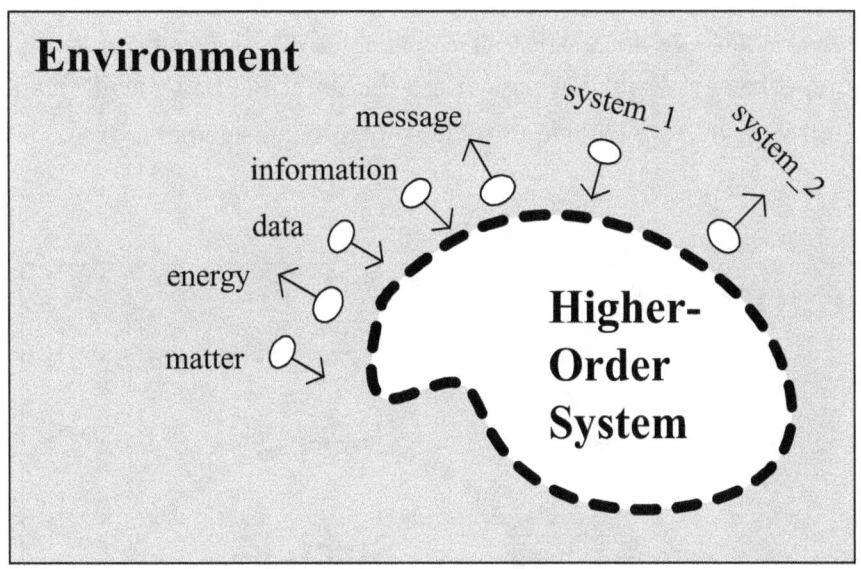

Figure 1-12   Higher-Order System

Human brain is regarded as higher-order systems. The human brain is a higher-order system, because it is able to produce a large number of systems, as shown in Figure 1-13. In the figure, *System_1*, *System_2*, and *System_n* are the output of the human brain.

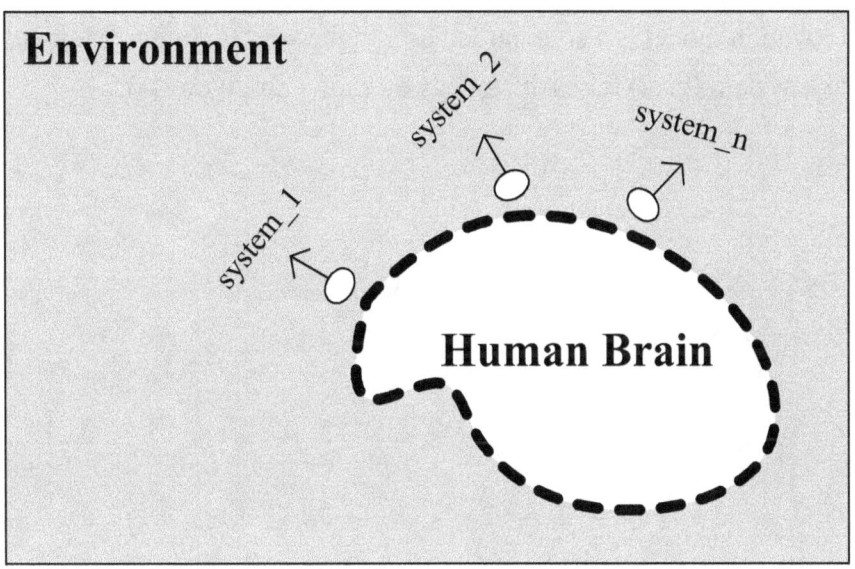

Figure 1-13   Human Brain is a Higher-Order System

Strategic management is also regarded as a higher-order system. Strategic management, for each strategy will output a system (goal), as shown in Figure 1-14. In the figure, *Strategy_1* and *Strategy_n* are the input of the strategic management; *System_1* and *System_n* are the output of the strategic management.

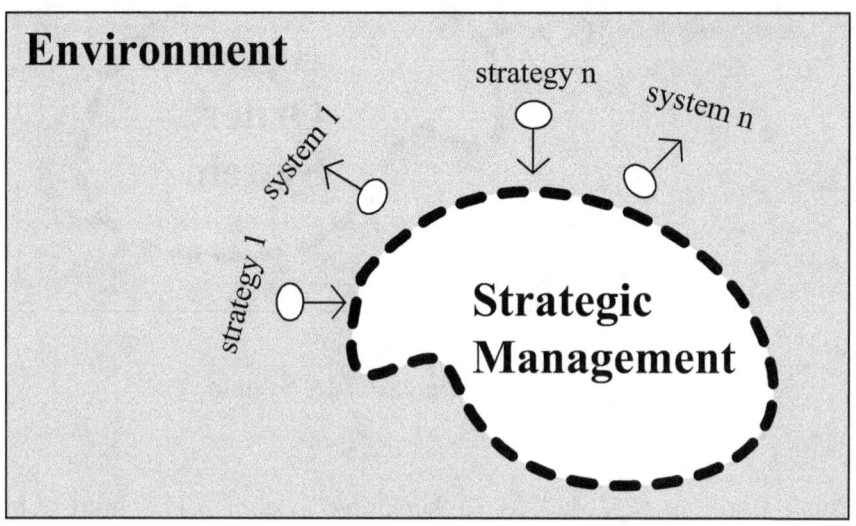

Figure 1-14   Strategic Management is a Higher-Order System

Motivation model is also regarded as a higher-order system. Motivation model will output a system (goal) for each strategy as shown in Figure 1-15.

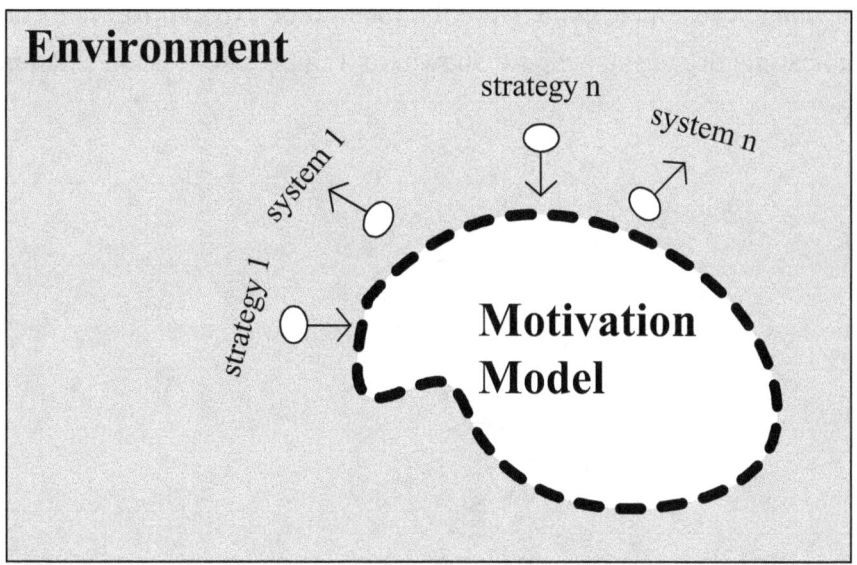

Figure 1-15   Motivation Model is a Higher-Order System

System dynamics is also regarded as a higher-order system, because it dynamically simulates the causal relationship among a large number of systems such as *System_1*, *System_2*, and *System_n* as shown in Figure 1-16. From these simulated systems, a decision maker thus is able to strategically choose the most appropriate one.

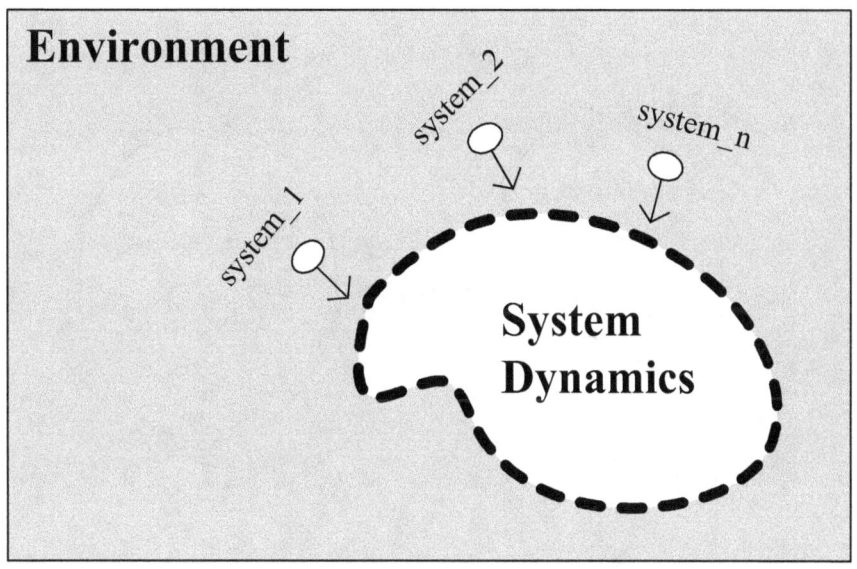

Figure 1-16   System Dynamics is a Higher-Order System

1-5 Evolution of a System

A system, not matter it is physical or virtual, will always change from time to time. The change cause may come from the internal or external forces of the system. A self-replicating organism cell, as shown in Figure 1-17, is an example of the internal forces.

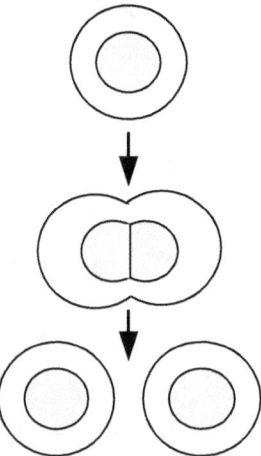

Figure 1-17   Self-Replicating Organism Cell

A worker reshaping, rebuilding, or remodeling a system, as shown in Figure 1-18, is an example of the external forces.

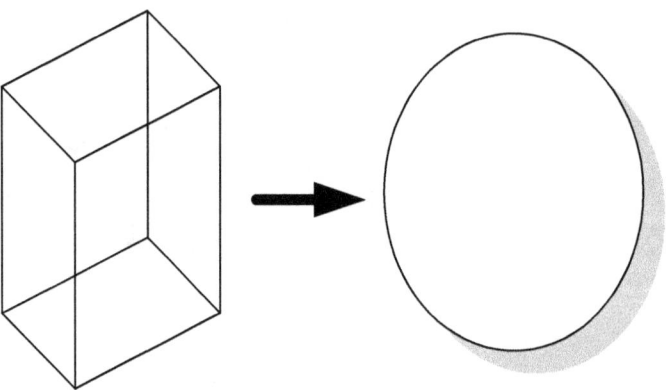

Figure 1-18   Reshape a System

A system evolves when it changes. Evolution of a system is shown in Figure 1-19.

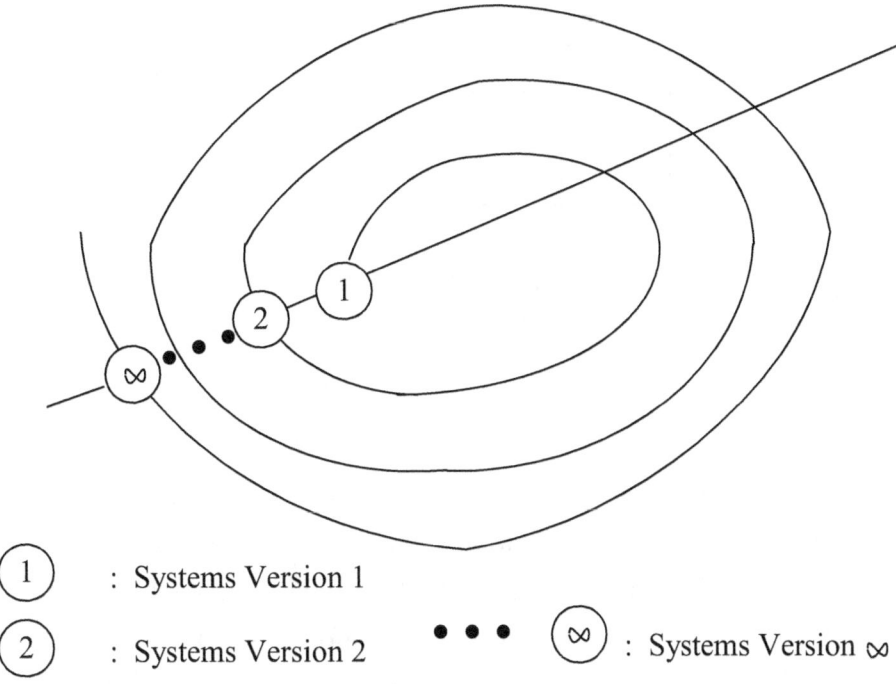

1   : Systems Version 1

2   : Systems Version 2   • • •   ∞   : Systems Version ∞

Figure 1-19   Evolution of a System

For example, Figure 1-20 shows the systems definition 1.0 systems definition *version 1* defining the *House_B* to be hopefully an integrated whole embodied in its assembled components of *Roof_1*, *Window_1*, and *Door_1*, their interrelationships with each other and the environment, and the principles and guidelines governing its design and evolution.

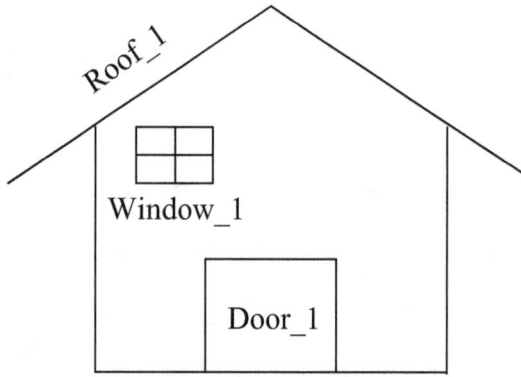

Figure 1-20   Systems Definition 1.0 Systems Definition *Version 1*
Defining the *House_B*

If the *house_B* changes and evolves for the first ime, Figure 1-21 shows the systems definition 1.0 systems definition *version 2* defining the *House_B* to be hopefully an integrated whole embodied in its assembled components of *Roof_1*, *Window_1*, *Window_2*, and *Door_1*, their interrelationships with each other and the environment, and the principles and guidelines governing its design and evolution.

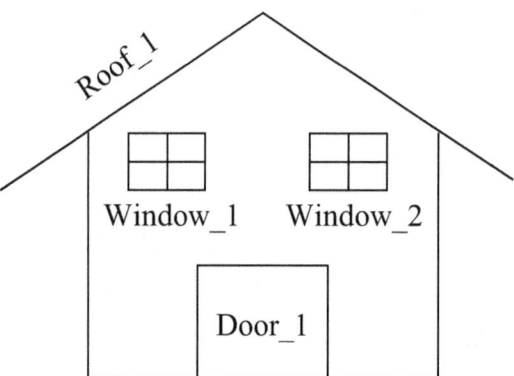

Figure 1-21   Systems Definition 1.0 Systems Definition *Version 2*
Defining the *House_B*

If the *house_B* changes and evolves a second ime, Figure 1-22 shows the systems definition 1.0 systems definition *version 3* defining the *House_B* to be hopefully an integrated whole embodied in its assembled components of *Roof_1*, *Window_1*, *Window_2*, *Door_1*, and *Door_2*, their interrelationships with each other and the environment, and the principles and guidelines governing its design and evolution.

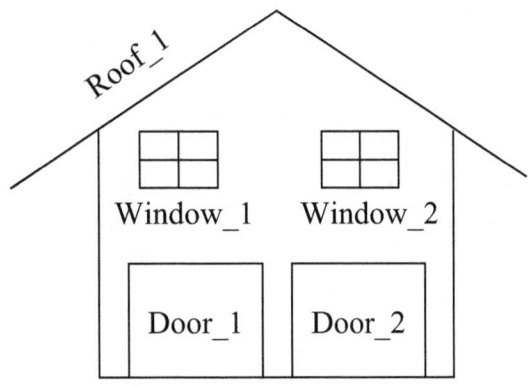

Figure 1-22   Systems Definition 1.0 Systems Definition *Version 3*
Defining the *House_B*

Chapter 2: Introduction to Systems Architecture

A system comprises multiple views such as strategy/version n, strategy/version n+1, concept, analysis, design, implementation, structure, behavior, and input/output data views. A systems model is required to describe and represent all these multiple views.

The systems model describes and represents the system multiple views possibly using two different approaches. The first one is the non-architectural approach and the second one is the architectural approach. The non-architectural approach respectively picks a model for each view. The architectural approach, instead of picking many heterogeneous and separated models, will use only one single coalescence model.

When used as a knowledge repository of a system, systems architecture becomes a communicating tool for comprehension enhancement, internal collaboration, and interworking with partners. Systems architecture also supplies documented systems structures and systems behaviors.

Systems architecture should not be constructed in one step. On the contrary, systems architects will iteratively and evolutionally construct each version of the systems architecture. Iterations and evolutions allow systems architects to demonstrate incremental value of their works and obtain early feedback of the systems architecture.

2-1 Multiple Views of a System

In general, a system is extremely complex that it consists of several evolution&motivation views such as strategy/version n and strategy/version n+1 views; it also consists of various multi-level (hierarchical) views such as concept, analysis, design, and implementation views; it also consists of many systemic views such as structure, behavior, and input/output data views [Date03, Elma10, Kend10, Pres09, Somm06].

Figure 2-1 shows that in a system all these strategy/version n, strategy/version n+1, concept, analysis, design, implementation, structure, behavior, and input/output data views represent the multiple views of a system.

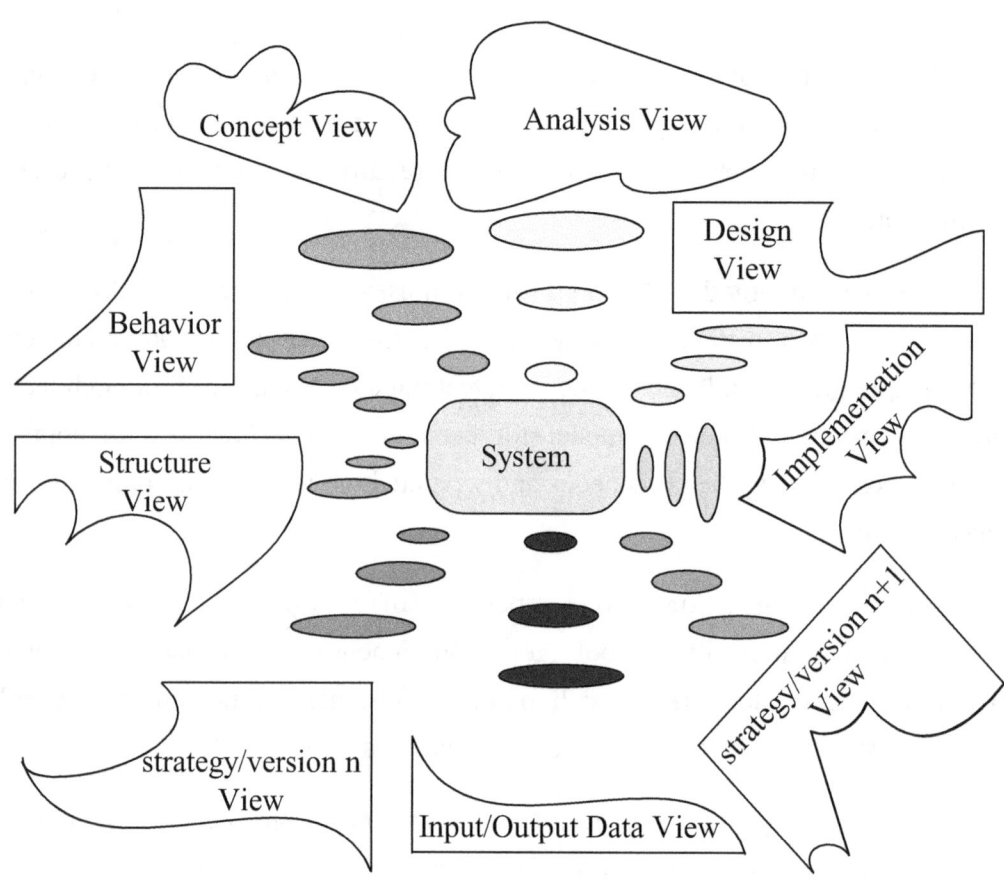

Figure 2-1   Multiple Views of a System

Among the above multiple views, the structure and behavior views are perceived as the two prominent ones. The structure view focuses on the systems structure which is described by components and their composition while the behavior view concentrates on the systems behavior which involves interactions (or handshakes)among the external environment's actors and components. Strategy/version n, strategy/version n+1, concept, analysis, design, implementation, and input/output data views are considered to be other views as shown in Figure 2-2.

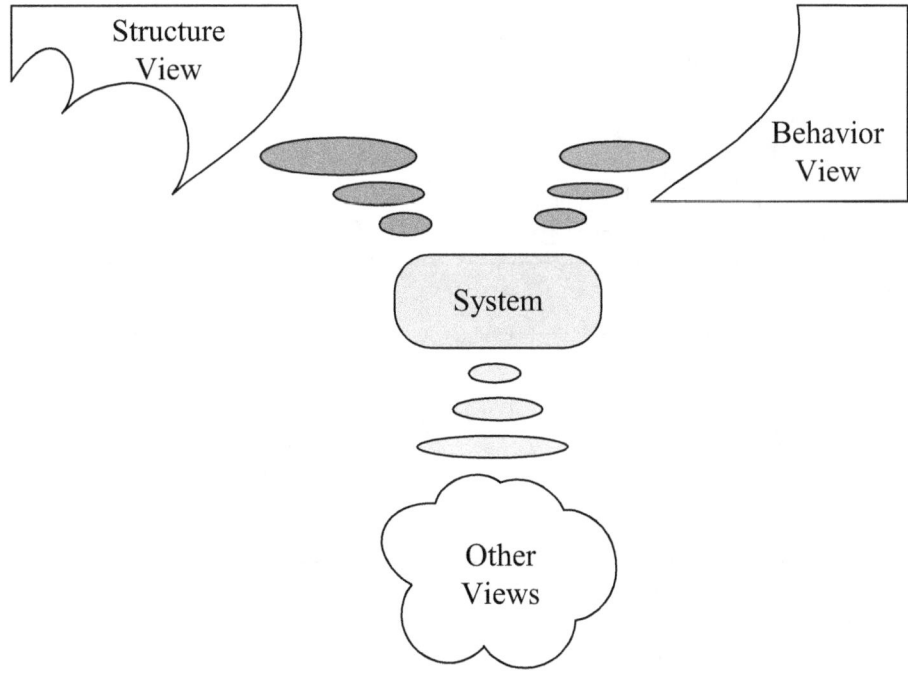

Figure 2-2   Structure, Behavior, and Other Views

Either Figure 2-1 or Figure 2-2 represents the multiple views of a system. In some situations Figure 2-1 is used and in other situations Figure 2-2 is used.

2-2 Systems Model

A systems model (SM) is a virtual system, distinguished from a physical system, used to describe and represent either the physical or virtual systems.

Figure 2-3 shows a physical system in which there are two buildings located in the upper left side and right underneath. The upper left building is Jackson Hotel and the right underneath building is Clinton Theater.

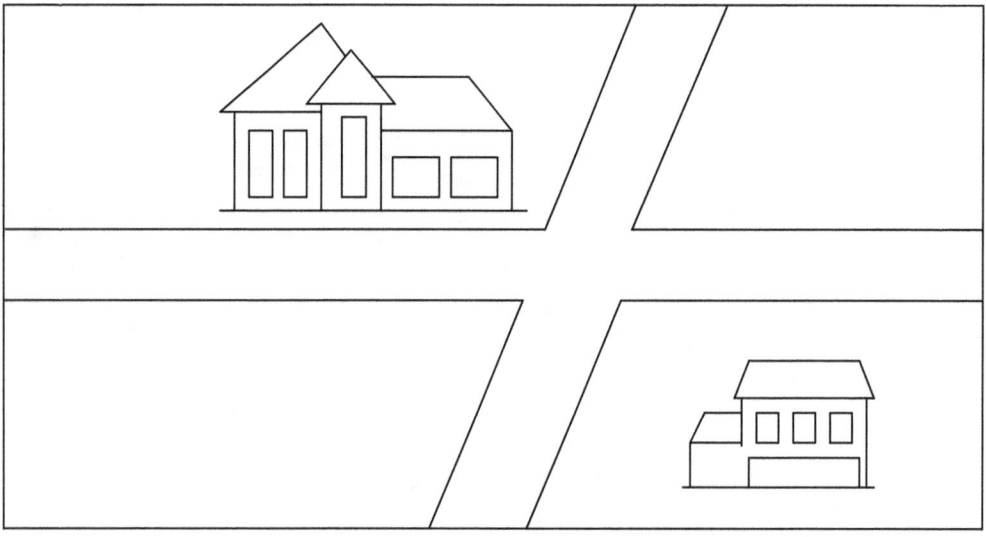

Figure 2-3   A Physical System

To model the physical system in Figure 2-3 we may then obtain a map as shown in Figure 2-4. The map is a kind of systems model used to describe and represent the physical system.

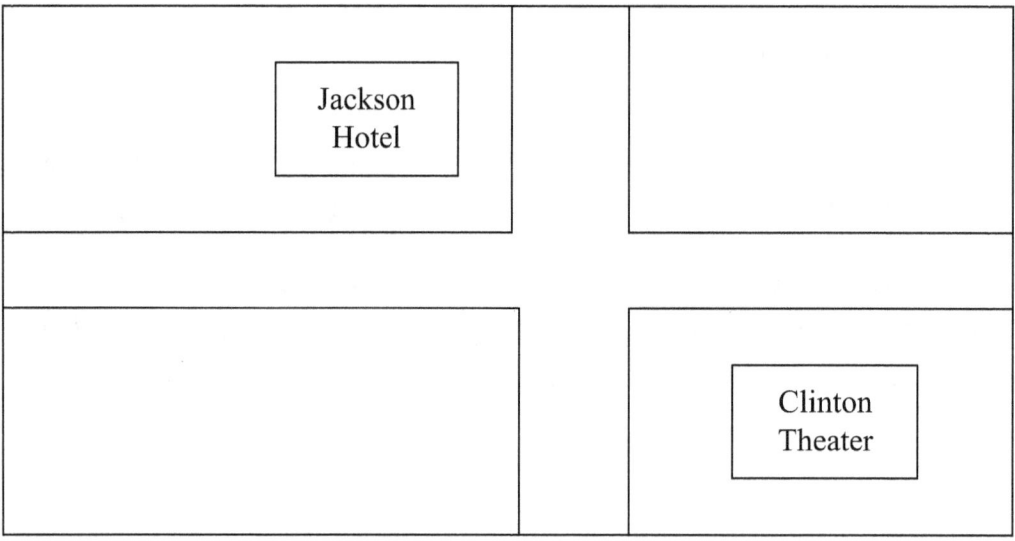

Figure 2-4   Map as a Systems Model

Besides describing and representing systems in the physical world, a systems model can also describe and represent systems in the virtual world. The virtual world includes a software system, a virtual reality, or a thought within a person's mind, etc. Figure 2-5 shows that a fashion designer is designing a new suit of clothes. Designing a suit of clothes, being a thought inside a person's mind, belongs to the virtual world.

Figure 2-5   Thought inside a Person's Mind

To model the thought within a person's mind in Figure 2-5, we may then use a clothes design diagram as shown in Figure 2-6. The clothes design diagram is a kind of systems model used to describe and represent a person's thought.

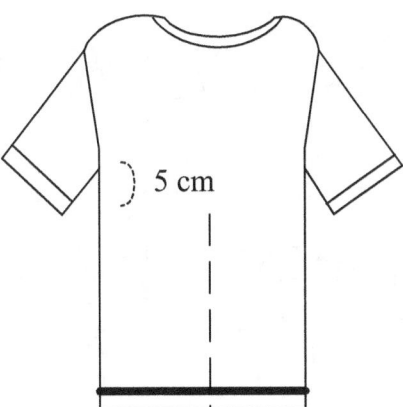

Figure 2-6   Clothes Design Diagram as a System Model

2-3 Non-Architectural Approaches Versus Architectural Approaches

A system is exceptionally complex that it includes multiple views such as strategy/version n, strategy/version n+1, concept, analysis, design, implementation, structure, behavior, and input/output data views.

The systems model describes and represents the system multiple views possibly using two different approaches. The first one is the non-architectural approach and the second one is the architectural approach.

The non-architectural approach, also known as the model multiplicity approach [Dori95, Dori02, Dori16], respectively picks a model for each view as shown in Figure 2-7, the strategy/version n view has the strategy/version n model, the strategy/version n+1 view has the strategy/version n+1 model, the concept view has the concept model, the analysis view has the analysis model, the design view has the design model, the implementation view has the implementation model, the structure view has the structure model, the behavior view has the behavior model, and the input/output data view has the input/output data model. These multiple models, are heterogeneous and not related to each other, and thus become the primary cause of model multiplicity problems [Dori95, Dori02, Dori16, Pele02, Sode03].

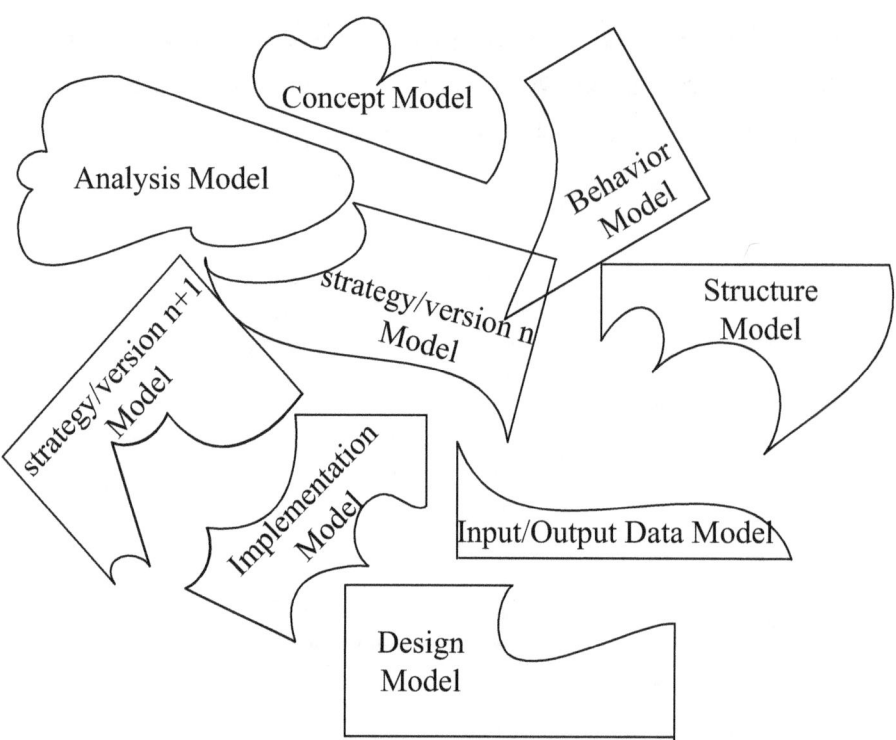

Figure 2-7   The Non-architectural Approach Picks a Model for Each View

The architectural approach, also known as the model singularity approach [Dori95, Dori02, Dori16, Pele02, Sode03], instead of picking many different models, will use only one single model as shown in Figure 2-8. The strategy/version n, strategy/version n+1, concept, analysis, design, implementation, structure, behavior, and input/output data views are all integrated in this multiple views coalescence (MVC) model of systems architecture [Chao14a, Chao14b, Chao14c, Clem02, Clem10, Dike01, Gort06, Putm00, Roza05, Shaw96, Tayl09, Wang99].

Figure 2-8   Systems Architecture Uses a Single Model

Figure 2-7 has many models. Figure 2-8 has only one model. Comparing Figure 2-7 with Figure 2-8, we unquestionably conclude that an integrated, holistic, united, coordinated, coherent, and coalescence model is more favorable than a collection of many heterogeneous and separated models.

2-4 Definition of Systems Architecture

Involved systems are extremely complex in every aspect so that each stakeholder needs a blueprint or model to capture their essential structures and behaviors. Systems architecture is such a blueprint or model.

There are several well-know definitions of systems architecture [Dam06, Mino08, O'Rou03, Roza05]. ANSI/IEEE 1471-2000 defines systems architecture as: "the fundamental organization of a system, embodied in its components, their relationships to each other and the environment, and the principles governing its design and evolution." The Open Group defines systems architecture as either "a formal description of a system, or a detailed plan of the system at component level to guide its implementation," or as "the structure of components, their interrelationships, and the principles and guidelines governing their design and evolution over time" [Rayn09, Toga08].

Concluding the above definitions, we now give systems architecture a definition of our own as shown in Figure 2-9.

Systems architecture is an integrated whole of a system's multiple views, i.e., structure, behavior, and other views, embodied in its components, their interactions with each other and the environment, and the principles and guidelines governing its design and evolution.

Figure 2-9   Definition of Systems Architecture

From the above definition, we find out that systems architecture is an integrated whole of a system's multiple views, i.e., structure, behavior, and other views, embodied in its assembled components, their interactions (or handshakes) with each other and the environment, and the principles and guidelines governing its design and evolution. That is, systems architecture is an integrated and coalescence model of multiple views. In this coalescence model, structure, behavior, and other views are all included in it as shown in Figure 2-10. We do not supply each view a respective model in this systems architecture coalescence model.

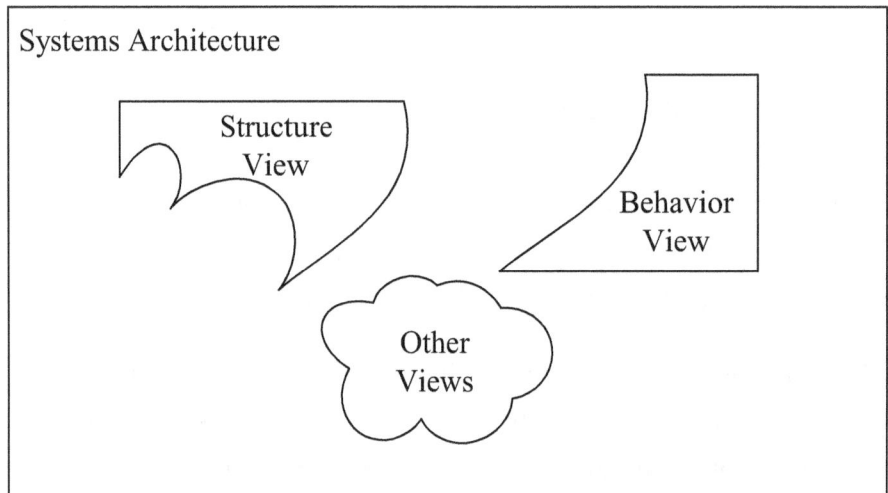

Figure 2-10   All Multiple Views are Included in This Systems Architecture

Since multiple views are embodied in a system's assembled components which belong to the structure view, they shall not exist alone. Multiple views must be loaded on the structure view just like a cargo is loaded on a ship as shown in Figure 2-11. There will be no multiple views if there is no structure view. Stand-alone multiple views are not meaningful.

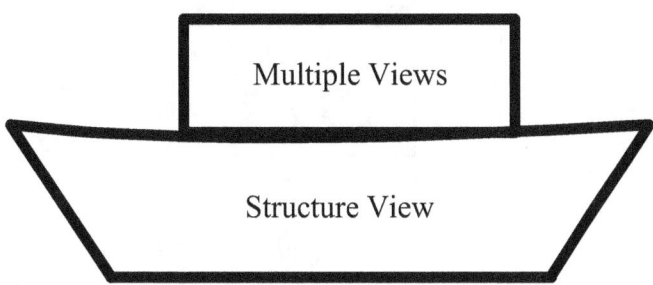

Figure 2-11   Multiple Views Must be Loaded on the Structure View

2-5 Systems Architecture as a Knowledge Repository

Based on its definition, systems architecture can be regarded as a knowledge repository of a system. Each stakeholder, through structure, behavior, and other views, submits his own knowledge and expertise to this repository when the systems architecture is built up, as shown in Figure 2-12.

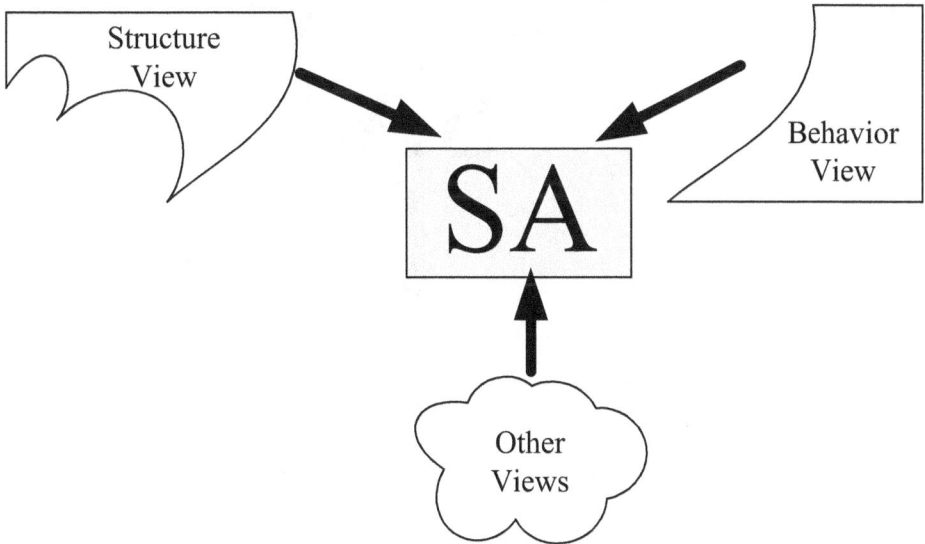

Figure 2-12   Each Stakeholder Submits His Own Knowledge and Expertise

On the other hand, any stakeholder, if there is any request then he would query the system architecture. The result of the query is gathered into a view for stakeholders to see or read, as shown in Figure 2-13.

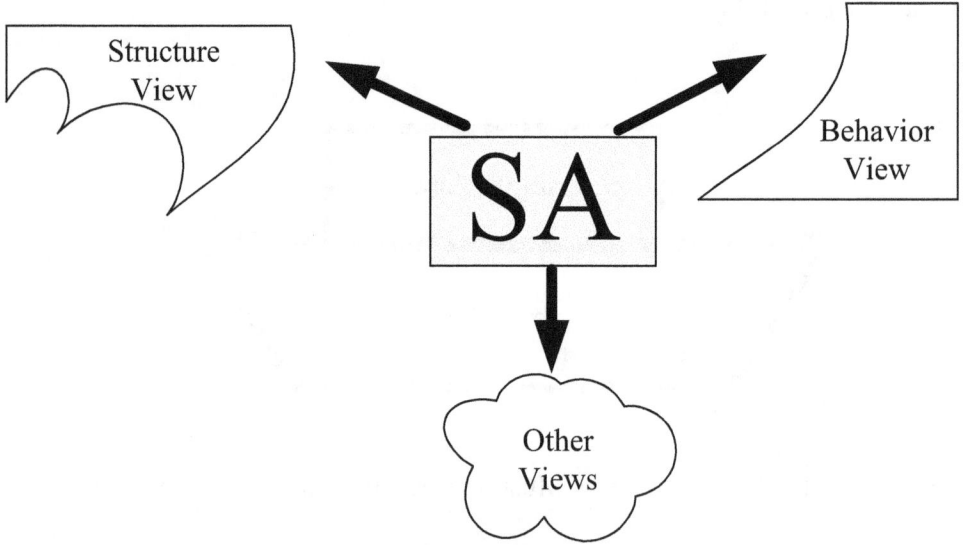

Figure 2-13   Views for Stakeholders to See or Read

Combining the above two figures, Figure 2-14 tells us that systems architecture is exactly a knowledge repository of a system. Stakeholders can submit and acquire knowledge to and from the systems architecture.

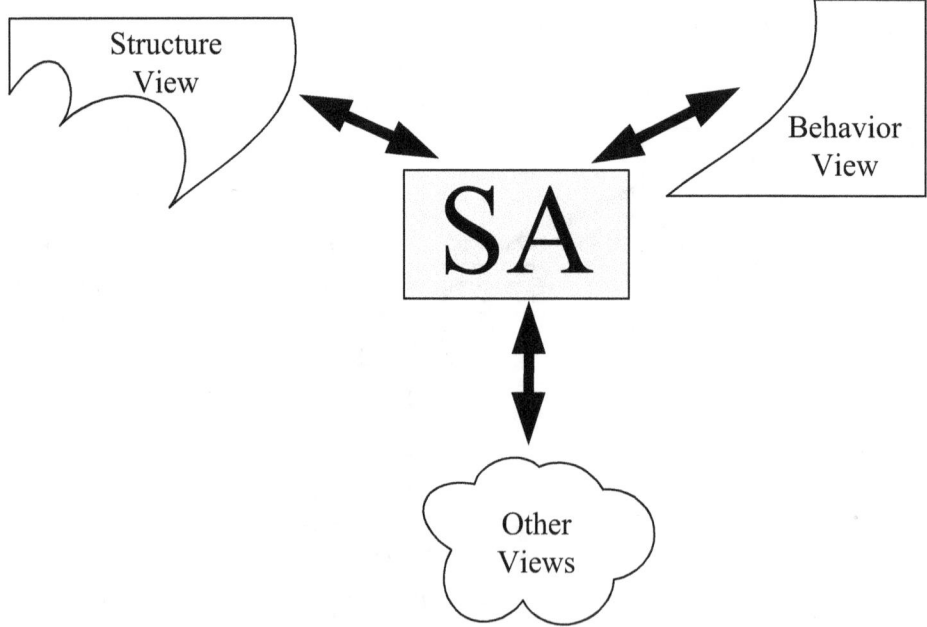

Figure 2-14   SA as a Knowledge Repository of a System

When used as a knowledge repository of a system, systems architecture becomes a communicating tool for comprehension enhancement, internal collaboration, and interworking with partners. The systems architecture also supplies documented systems structures and systems behaviors.

2-6 Constructing the Systems Architecture Iteratively and Evolutionally

Systems architecture shall not be constructed in one step. On the contrary, a systems architect must construct the systems architecture iteratively and evolutionally. Iterations and evolutions allow systems architects to demonstrate incremental values of their works and obtain early feedback of the systems architecture.

Figure 2-15 shows that the systems architecture *version 1*, *version 2*, *version 3*, *version 4*,…, and *version* ∞ are constructed iteratively and evolutionally by a systems architect.

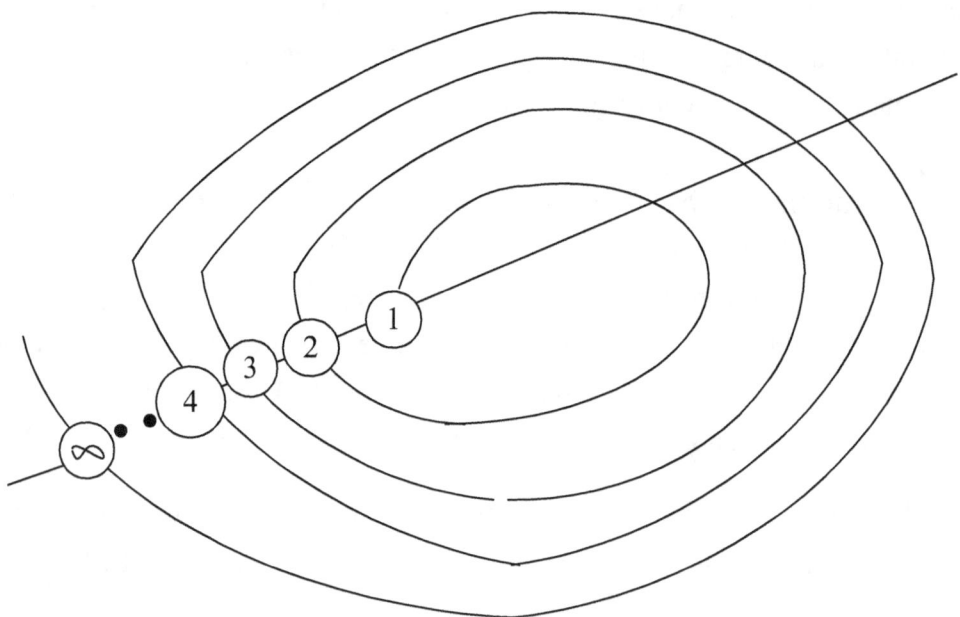

(1) : Systems Architecture Version 1

(2) : Systems Architecture Version 2

(3) : Systems Architecture Version 3

(4) : Systems Architecture Version 4

• • •

(∞) : Systems Architecture Version ∞

Figure 2-15   Systems Architecture is Constructed Iteratively and Evolutionally

Systems architecture *version n* is sometimes referred to as the baseline (As-Is) architecture which represents the current system that has been formally reviewed and agreed upon. On the other hand, systems architecture *version n+1* is sometimes referred to as the target (To-Be) architecture which represents the goal system that will be formally constructed.

2-7 View Model

A system comprises multiple views such as strategy/version n, strategy/version n+1, concept, analysis, design, implementation, structure, behavior, and input/output data views. We can represent all these multiple views in a one-dimensional array as shown in Figure 2-16.

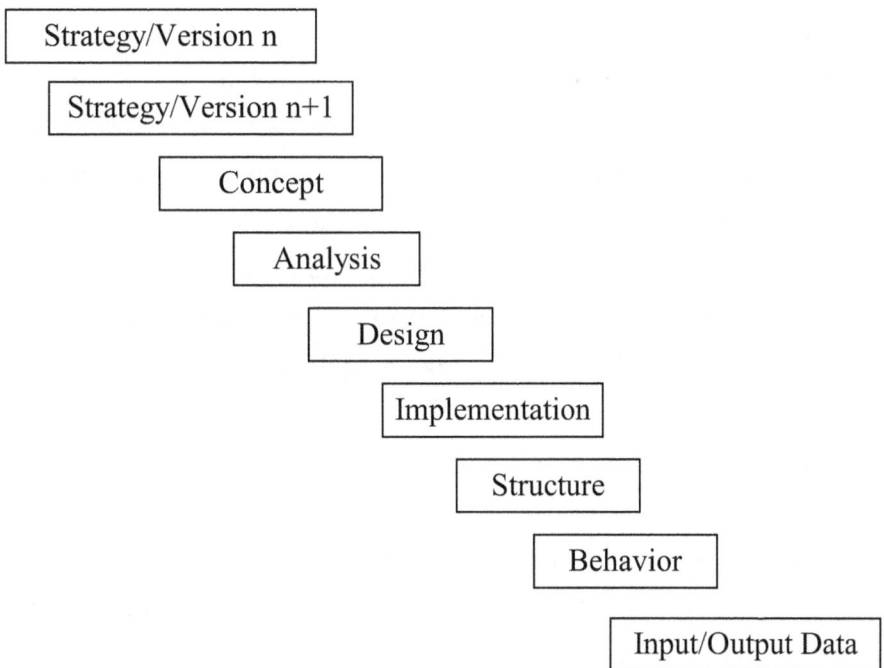

Figure 2-16   Array Representation of Multiple Views

We can also describe and represent all these multiple views in a three-dimensional matrix as shown in Figure 2-17.

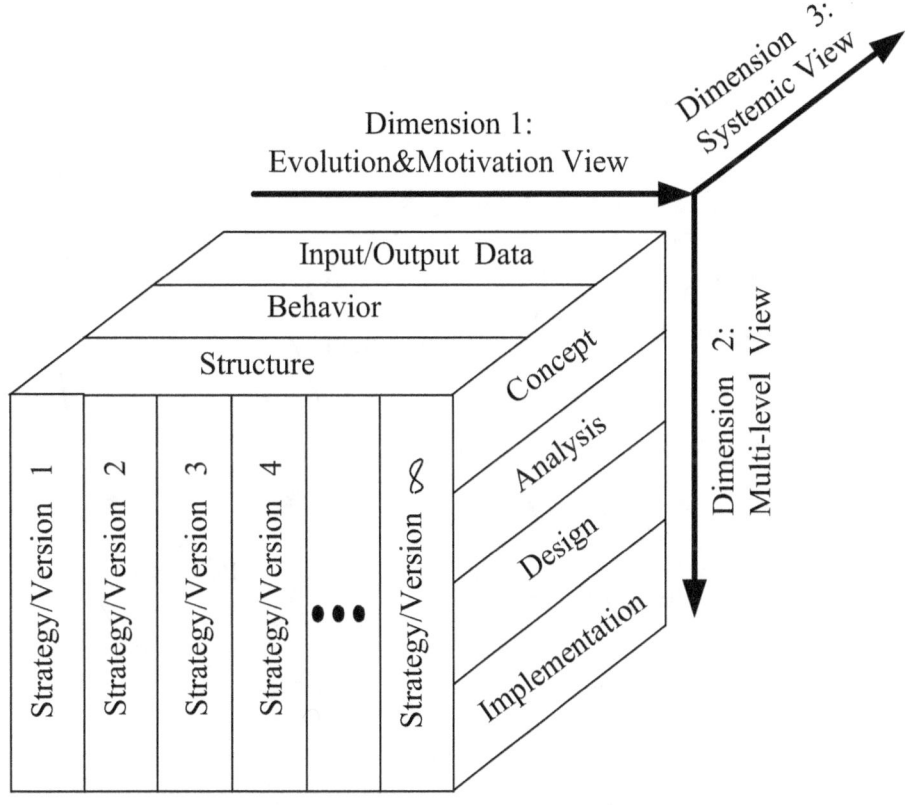

Figure 2-17   Three-dimensional Matrix Representation of Multiple Views

In the matrix representation of multiple views, dimension 1 stands for the evolution&motivation view which contains the strategy/version 1, strategy/version 2, strategy/version 3, strategy/version 4,…, and strategy/version ∞ views; dimension 2 stands for the multi-level (hierarchical) view which contains the concept, analysis, design, and implementation views; dimension 3 stands for the systemic view which contains the structure, behavior, input/output data views. The matrix representation of multiple views is also called a view model (VM) or architecture framework (AF) [Chao14a, Chao14b, Chao14c, Dam06, Mino08, O'Rou03].

2-8 Architecture Development Method

If we adopt the iterative and evolutional construction of systems architecture approach, then we would obtain the architecture development method (ADM) [Toga08] as shown in Figure 2-18.

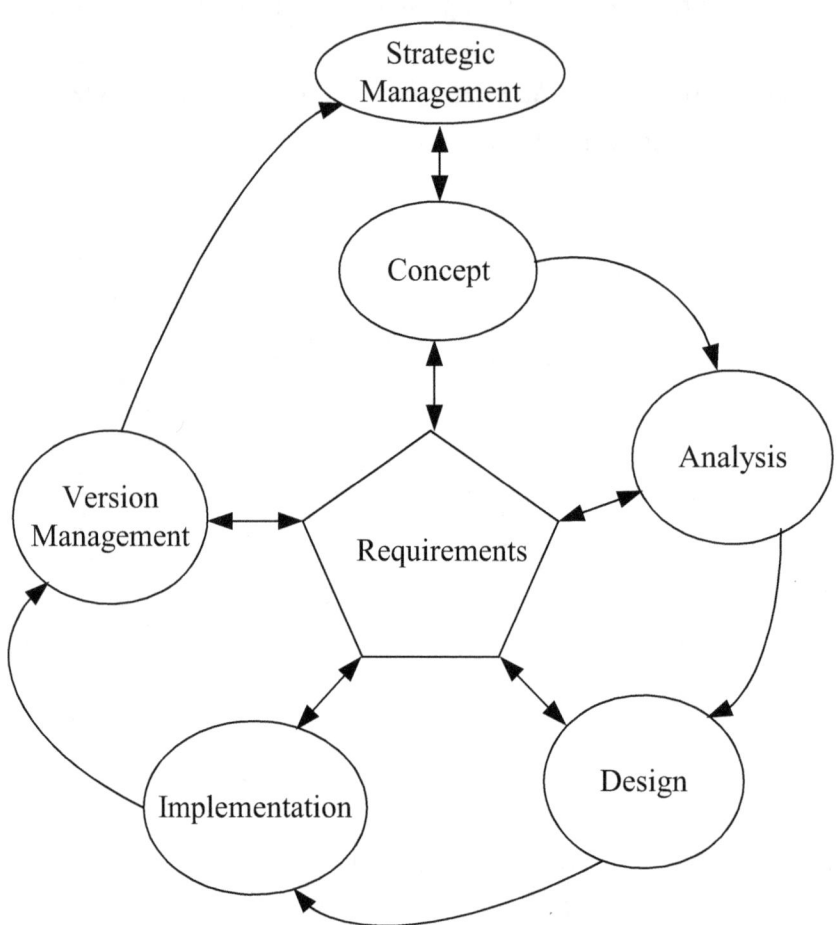

Figure 2-18   Architecture Development Method

The iterative and cyclic ADM, being utilized by a systems architect to accomplish each version management of systems architecture, shall do the strategic management first and then go through the concept, analysis, design, and implementation phases of systems architecture construction. Every phase checks with the requirements to make sure that each version of the constructed systems architecture is what the users want.

The output of strategic management is a strategy and the output of version management is a version of systems architecture. Accordingly, each strategy is mapped to a version of systems architecture as shown in Figure 2-19.

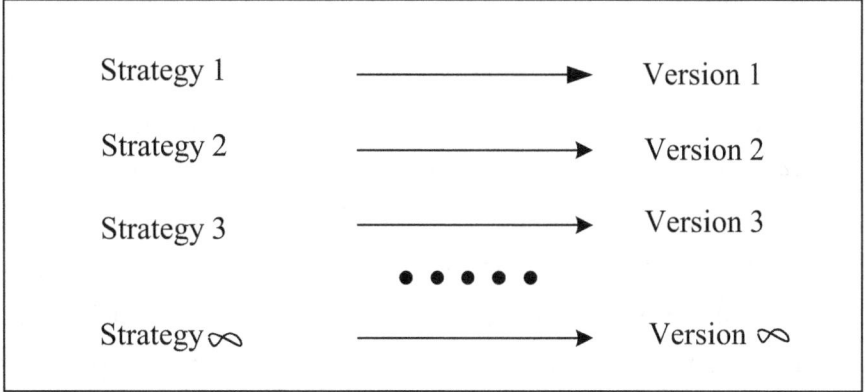

Figure 2-19   Each Strategy is Mapped to a Systems Architecture Version

2-9 Process Algebra as the Architecture Description Language

An architecture description is a formal description and representation of a system. A description of the systems architecture has to grasp the essence of the system and its details at the same time. In other words, an architecture description not only provides an overall picture that summarizes the whole system, but also contains enough detail that the system can be constructed and validated.

The language for architecture description is called the architecture description language (ADL) [Clem02, Clem10, Dike01, Roza05, Shaw96, Tayl09]. An ADL is a special kind of language used in describing the architecture of a system.

Process algebras are a diverse family of related approaches to the study of concurrent systems [Berg87, Hoar85, Miln89, Miln99]. Their tools are algebraic languages for the high-level description of interactions, communications, and synchronizations among independent processes. As "interaction" plays an important factor in integrating the systems structure and systems behavior for a system, it is very appropriate that we use a process algebra approach as the architecture description language.

Since the architectural approach uses a coalescence model for all multiple views of a system, the foremost duty of this process algebra approach as the architecture description language is to make the strategy/version n, strategy/version n+1, concept, analysis, design, implementation, structure, behavior, and input/output data views all integrated and coalesced within this architecture description.

In general, we shall use process algebra to describe the three-dimensional matrix representation of multiple views as shown in Figure 2-20. That is, process algebra will describe the systemic view (dimension 3) which contains the structure, behavior, input/output data views, and also describe the multi-level view (dimension 2) which contains the concept, analysis, design, and implementation views, and last describe the evolution&motivation view (dimension 1) which contains the strategy/version 1, strategy/version 2, strategy/version 3, strategy/version 4,…, and strategy/version ∞ views.

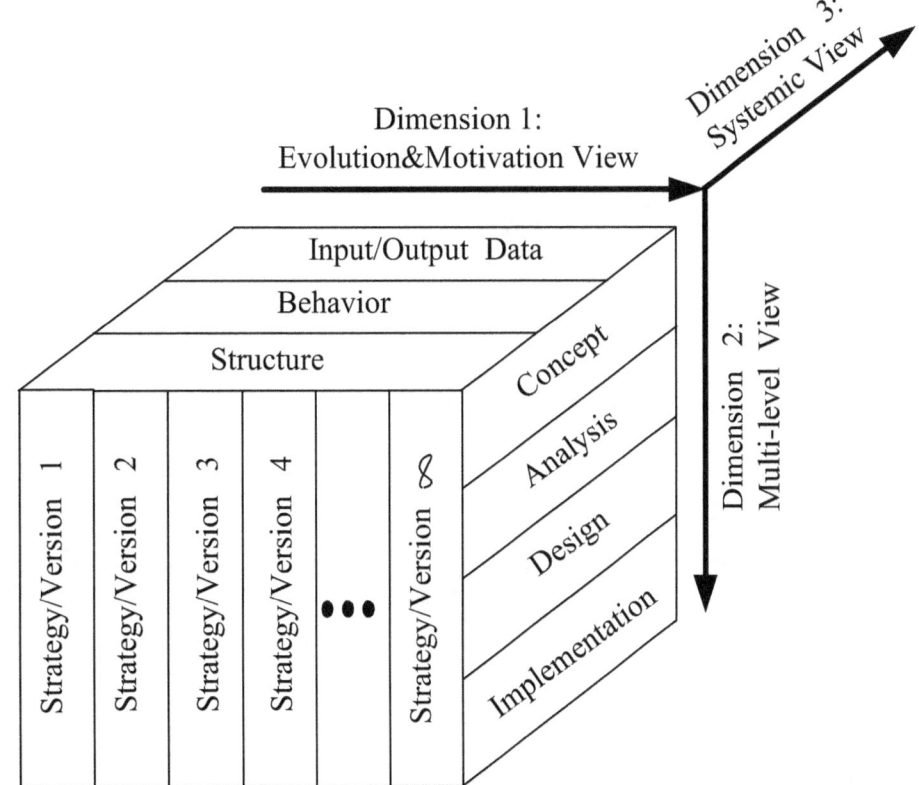

Figure 2-20   Process Algebra to Describe the Three-Dimensional Matrix
Representation of Multiple Views

Chapter 3: Structure-Behavior Coalescence for Systems Architecture

In general, multiple views coalescence (MVC) architecture is synonymous with the systems architecture. Since structure and behavior views are the two most prominent ones among multiple views, integrating the structure and behavior views becomes a superb approach for integrating multiple views of a system. In other words, structure-behavior coalescence (SBC) leads to the coalescence of multiple views. Therefore, we conclude that SBC architecture is also synonymous with the systems architecture.

3-1 Multiple Views Coalescence to Achieve the Systems Architecture

Systems architecture has been defined as a coalescence model of multiple views. Multiple views coalescence uses only a single coalescence model as shown in Figure 3-1. strategy/version n, strategy/version n+1, concept, analysis, design, implementation, structure, behavior, and input/output data views are all integrated in this MVC architecture.

Figure 3-1   MVC Architecture

Generally, MVC architecture is synonymous with the systems architecture. In other words, multiple views coalescence sets a path to achieve the systems architecture as shown in Figure 3-2.

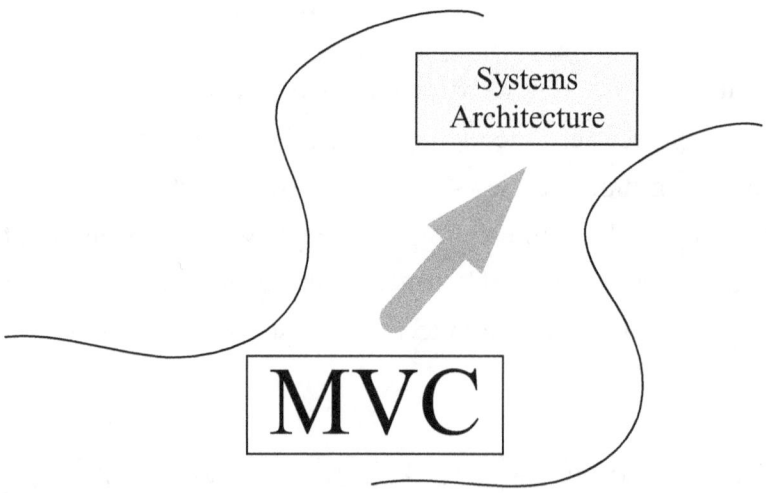

Figure 3-2   MVC to Achieve the Systems Architecture

In the MVC architecture, multiple views must be attached to or built on the systems structure. In other words, multiple views shall not exist alone; they must be loaded on the systems structure just like a cargo is loaded on a ship as shown in Figure 3-3. There will be no multiple views if there is no systems structure. Stand-alone multiple views are not meaningful.

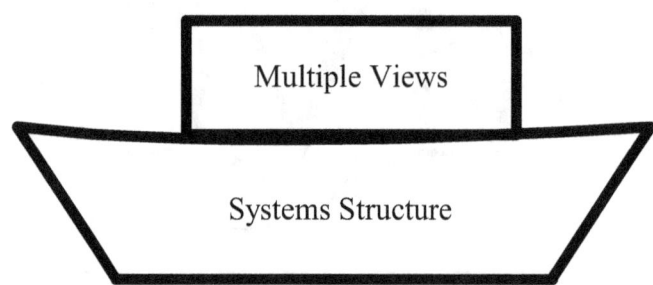

Figure 3-3   Multiple Views are Loaded on the Systems Structure

3-2 Integrating the Systems Structures and Systems Behaviors

By integrating the systems structure and systems behavior, we obtain structure-behavior coalescence (SBC) within the system as shown in Figure 3-4.

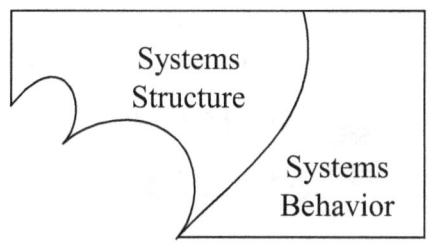

Figure 3-4   Structure-Behavior Coalescence

Structure-behavior coalescence has never been used in any systems model (SM) for systems development except the SBC architecture. There are many advantages to use the structure-behavior coalescence approach to integrate the systems structure and systems behavior.

SBC architecture uses a single model as shown in Figure 3-5. Systems structures and systems behaviors are integrated in this SBC architecture.

Figure 3-5   SBC Architecture

Since systems structures and systems behaviors are so tightly integrated, we sometimes claim that the core theme of SBC architecture is: "Systems Architecture = Systems Structure + Systems Behavior," as shown in Figure 3-6.

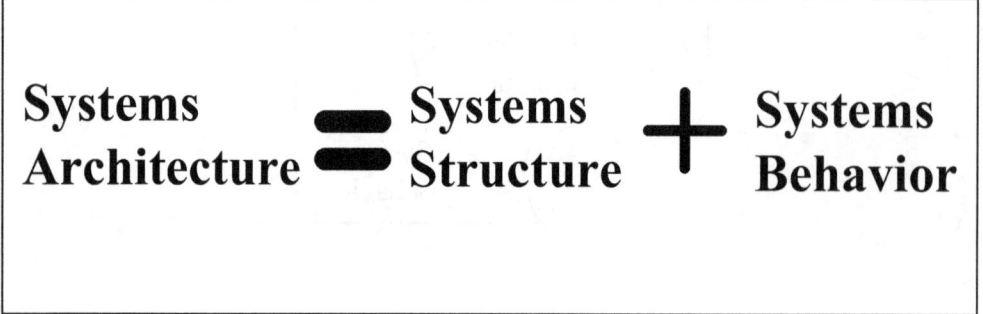

Figure 3-6   Core Theme of SBC Architecture

So far, systems behaviors are separated from systems structures in most cases [Pres09, Roza05, Somm06]. For example, the well-known structured systems analysis and design (SSA&D) approach uses structure charts (SC) to represent the systems structure and data flow diagrams (DFD) to represent the systems behavior [Denn08, Kend10, Your99]. SC and DFD are two different models. They are so separated like that there is "Pacific Ocean" between them, as shown in Figure 3-7.

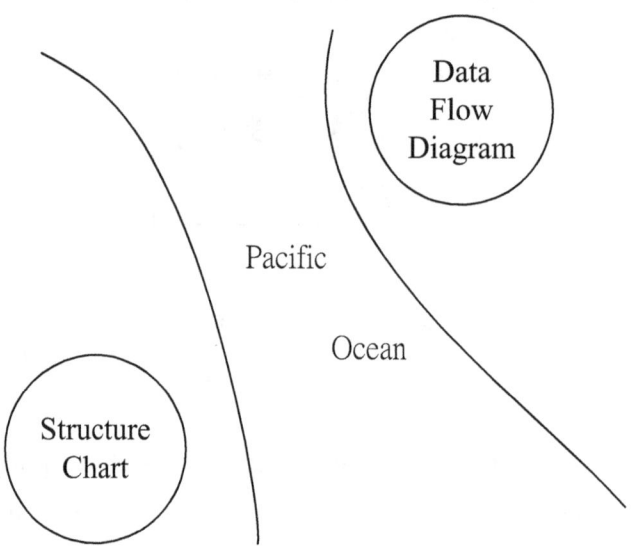

Figure 3-7   Two Heterogeneous and Separated Models

3-3 Structure-Behavior Coalescence to Facilitate Multiple Views Coalescence

Since structure and behavior views are the two most prominent ones among multiple views, integrating the structure and behavior views is clearly the best way to integrate multiple views of a system. In other words, structure-behavior coalescence facilitates multiple views coalescence as shown in Figure 3-8.

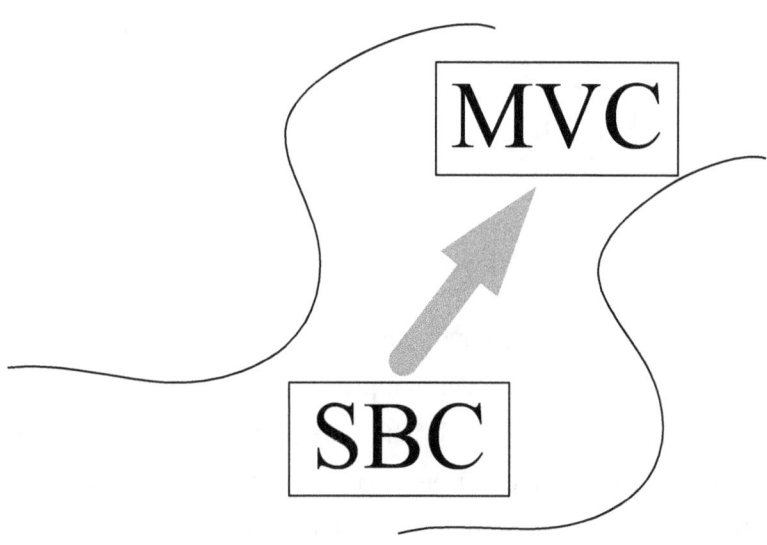

Figure 3-8   SBC Facilitates MVC

3-4 Structure-Behavior Coalescence to Achieve the Systems Architecture

Figure 3-2 declares that multiple views coalescence sets a path to achieve the desired systems architecture with the most efficient approach. Figure 3-8 declares that structure-behavior coalescence facilitates multiple views coalescence.

Combining the above two declarations, we conclude that structure-behavior coalescence sets a path to achieve the systems architecture as shown in Figure 3-9. In this case, SBC architecture is also synonymous with the systems architecture.

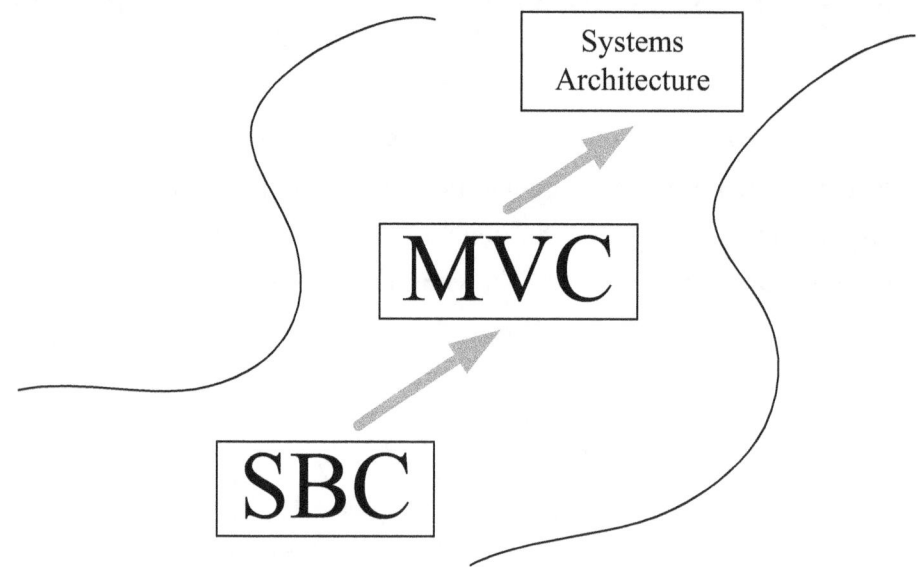

Figure 3-9   SBC to Achieve the Systems Architecture

SBC architecture strongly demands that the structure and behavior views must be coalesced and integrated. This never happens in other architectural approaches such as Zachman Framework [O'Rou03], The Open Group Architecture Framework (TOGAF) [Rayn09, Toga08], Department of Defense Architecture Framework (DoDAF) [Dam06] and Unified Modeling Language (UML) [Rumb91]. Zachman Framework does not offer any mechanism to integrate the structure and behavior views. TOGAF, DoDAF and UML do not, either.

In the SBC architecture, a systems behavior must be attached to or built on a systems structure. In other words, a systems behavior can not exist alone; it must be loaded on a systems structure just like a cargo is loaded on a ship as shown in Figure 3-10. There will be no systems behavior if there is no systems structure. A stand-alone systems behavior is not meaningful.

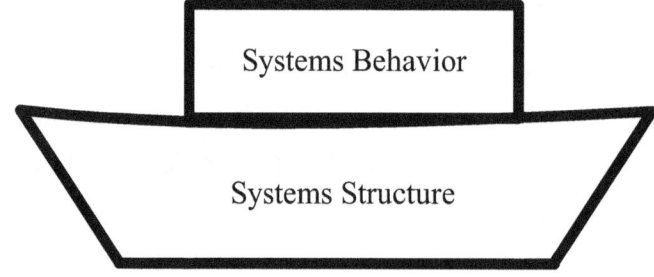

Figure 3-10   A Systems Behavior Must be Loaded on a Systems Structure

Chapter 4: Introduction to SBC Architecture

As discussed in the previous chapter, SBC (i.e. structure-behavior coalescence) architecture is the recommended systems architecture approach. SBC architecture includes: a) SBC view model (SBC-VM) , b) SBC architecture development method (SBC-ADM) , and c) multi-queue SBC process algebra (M-SBC-PA).

View model (VM) is a three-dimensional matrix representation of multiple views of a system. In the SBC view model, dimension 1 stands for the evolution&motivation view which contains the strategy/version 1, strategy/version 2, strategy/version 3, strategy/version 4, strategy/version ∞ views; dimension 2 stands for the multi-level (hierarchical) view which contains the concept, analysis, design, and implementation views; dimension 3 stands for the systemic view which contains the collection of all interaction flow diagrams.

Through the iterative and cyclic architecture development method (ADM), a systems architect is able to construct the systems architecture smoothly. SBC-ADM, being utilized by a systems architect to accomplish each version management of the systems architecture, shall do the strategic management first and then go through the concept, analysis, design, and implementation phases of systems architecture construction. Every phase checks with the requirements to make sure that each version of the constructed systems architecture is what the users want.

Multi-queue SBC process algebra is a kind of architecture description language (ADL). Since the architectural approach uses a coalescence model for all multiple views of a system, the foremost duty of this multi-queue SBC process algebra as the architecture description language is to make the strategy/version n, strategy/version n+1, concept, analysis, design, implementation, collection of all interaction flow diagrams views all integrated and coalesced within this architecture description.

4-1 Definition of SBC Architecture

Here, let us first give the SBC architecture a definition as shown in Figure 4-1.

SBC architecture,
through structure-behavior coalescence,
truly is an integrated whole of a system's multiple views, i.e., structure, behavior, and other views, embodied in its components, their interactions with each other and the environment, and the principles and guidelines governing its design and evolution.

Figure 4-1   Definition of SBC Architecture

From the above definition, we find out that SBC architecture, through structure-behavior coalescence, is a truly integrated whole of a system's multiple views, i.e., structure, behavior, and other views, embodied in its assembled components, their interactions (or handshakes) with each other and the environment, and the principles and guidelines governing its design and evolution.

SBC architecture includes: a) SBC view model (SBC-VM), b) SBC architecture development method (SBC-ADM), and c) multi-queue SBC process algebra (M-SBC-PA).

4-2 SBC View Model

View model [Chao14a, Chao14b, Chao14c, Dam06, Mino08, O'Rou03] is a three-dimensional matrix representation of a system's multiple views as shown in Figure 4-2. In the figure, dimension 1 stands for the evolution&motivation view which contains the strategy/version 1, strategy/version 2, strategy/version 3, strategy/version 4,…, and strategy/version ∞ views; dimension 2 stands for the multi-level (hierarchical) view which contains the concept, analysis, design, and implementation views; dimension 3 stands for the systemic view which contains the structure, behavior, and input/output data views.

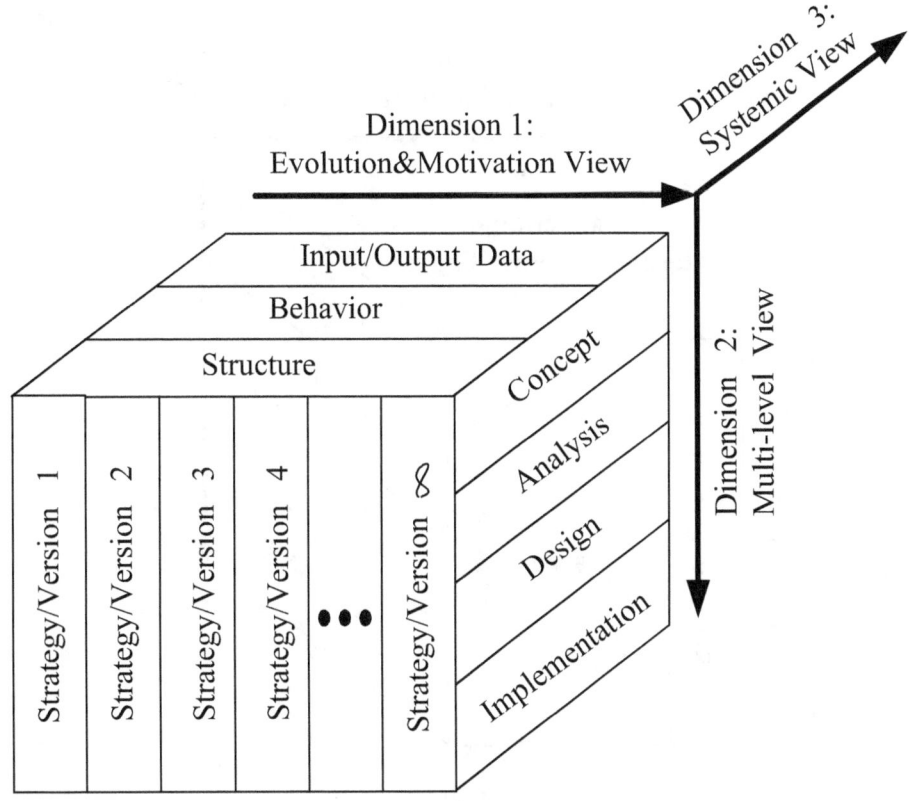

Figure 4-2   View Model

According to the SBC approach, the collection of all interaction flow diagrams defines the integration of the structure, behavior, and input/output data views. Adding these ideas to Figure 4-2, we then get the SBC view model as shown in Figure 4-3.

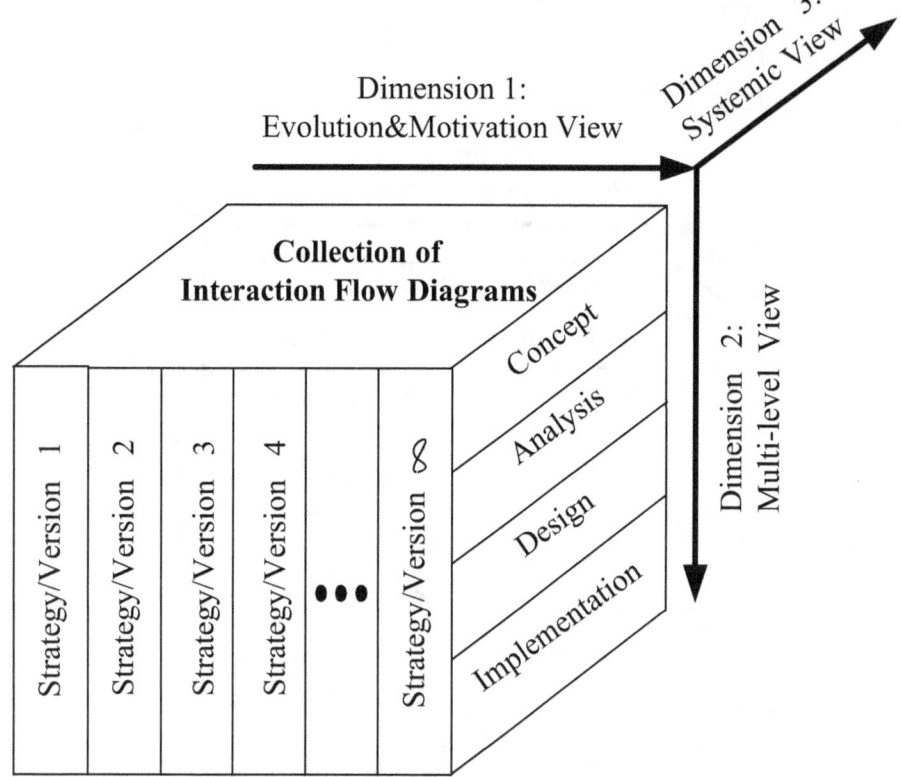

Figure 4-3   SBC View Model

In the SBC view model, dimension 1 stands for the evolution&motivation view which contains the strategy/version 1, strategy/version 2, strategy/version 3, strategy/version 4,…, and strategy/version ∞ views; dimension 2 stands for the multi-level (hierarchical) view which contains the concept, analysis, design, and implementation views; dimension 3 stands for the systemic view which contains the collection of all interaction flow diagrams (IFD).

4-3 SBC Architecture Development Method

The term of architecture development method (ADM) is first used in the open group architecture framework [Toga08]. Through the iterative and cyclic ADM, a systems architect is able to construct the systems architecture smoothly.

SBC architecture development method (SBC-ADM), being utilized by a systems architect to accomplish each version management of the systems architecture, shall do the strategic management first and then go through the concept, analysis, design, and implementation phases of systems architecture construction. Every phase shall check with the requirements to make sure that each version of the constructed systems architecture is what the users want as shown in Figure 4-4.

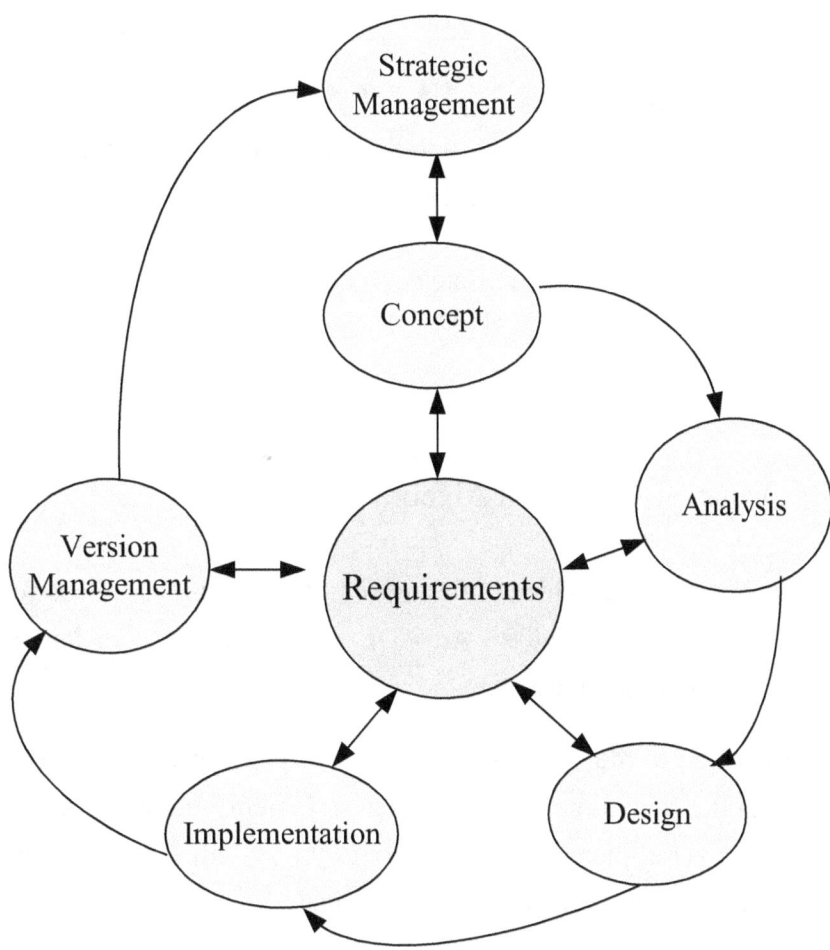

Figure 4-4   SBC Architecture Development Method

The output of strategic management is a strategy and the output of version management is a version of systems architecture. Accordingly, each strategy is mapped to a version of systems architecture as shown in Figure 4-5.

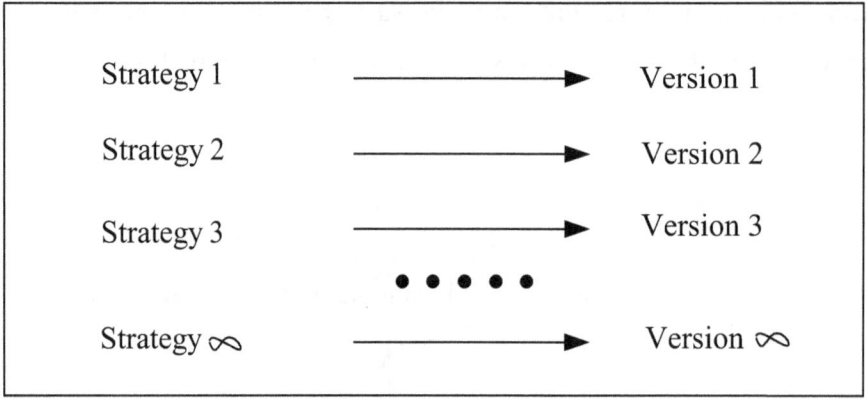

Figure 4-5   Each Strategy is Mapped to a Version of
Systems Architecture

4-4 Multi-Queue SBC Process Algebra

The language for architecture description is called the architecture description language (ADL) [Shaw96, Tayl09]. An ADL is a special kind of language used in describing the architecture of a system.

A description of the systems architecture has to grasp the essence of a system and its details at the same time. In other words, a systems architecture description not only provides an overall picture that summarizes the system, but also contains enough detail that the system can be constructed and validated.

Process algebras are a diverse family of related approaches to the study of concurrent systems [Berg87, Hoar85, Miln89, Miln99]. Their tools are algebraic languages for the high-level description of interactions, communications, and synchronizations among independent processes. As "interaction" plays an important factor in integrating the systems structure and systems behavior for a system, it is very appropriate that we use a process algebra approach as the architecture description language.

In the structure-behavior coalescence (SBC) approach, we shall use multi-queue SBC process algebra (M-SBC-PA) [Chao15f, Chao15h] as the architecture description language.

Since the architectural approach uses a coalescence model for all multiple views of a system, the foremost duty of this multi-queue SBC process algebra as the architecture description language is to make the strategy/version n, strategy/version n+1, concept, analysis, design, implementation, structure, and behavior views all integrated and coalesced within this architecture description.

In general, multi-queue SBC process algebra will describe the three-dimensional matrix representation of multiple views as shown in Figure 4-6. That is, multi-queue SBC process algebra will describe the systemic view (dimension 3) which contains the collection of all interaction flow diagrams, and also describe the multi-level view (dimension 2) which contains the concept, analysis, design, and implementation views, and last describe the evolution&motivation view (dimension 1) which contains the strategy/version 1, strategy/version 2, strategy/version 3, strategy/version 4,…, and strategy/version ∞ views.

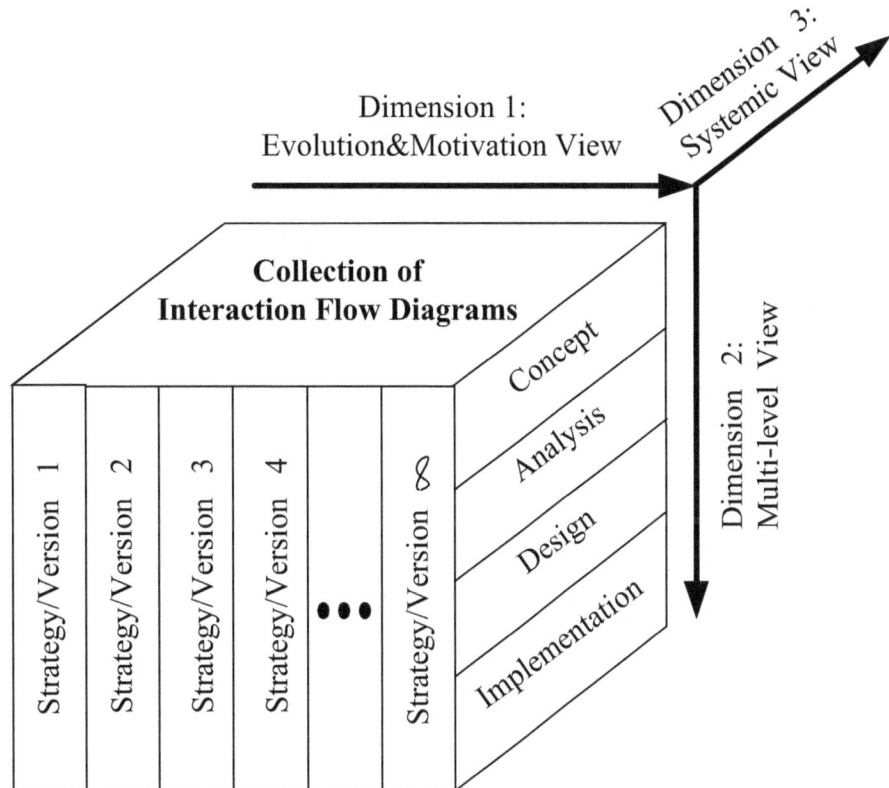

Figure 4-6   Multi-Queue SBC Process Algebra to Describe the Three-Dimensional Matrix Representation of Multiple Views

Chapter 5: Introduction to Systems Definition 2.0

Because systems definition 1.0 defining a system does not describe the integration of systems structure and systems behavior, very likely it will never be able to form a truly integrated whole of a system. In this situation, systems definition 1.0 is powerless in defining a system appropriately.

SBC architecture provides an elegant way to integrate the structure and behavior of a system. Therefore, systems definition 2.0 shall use the SBC architecture to define a system. A system is redefined, by systems definition 2.0, shown in Figure 5-1.

A system,
through the SBC architecture,
truly is an integrated whole,
embodied in its components,
their interactions with each other and the environment,
and the principles and guidelines governing its design and evolution.

Figure 5-1 Systems Definition 2.0 Defining a System

According to the above definition, systems definition 2.0 uses the SBC architecture to define a system. Based on the view model shown in Figure 5-2, there are three significant dimensions for systems definition 2.0 to define a system.

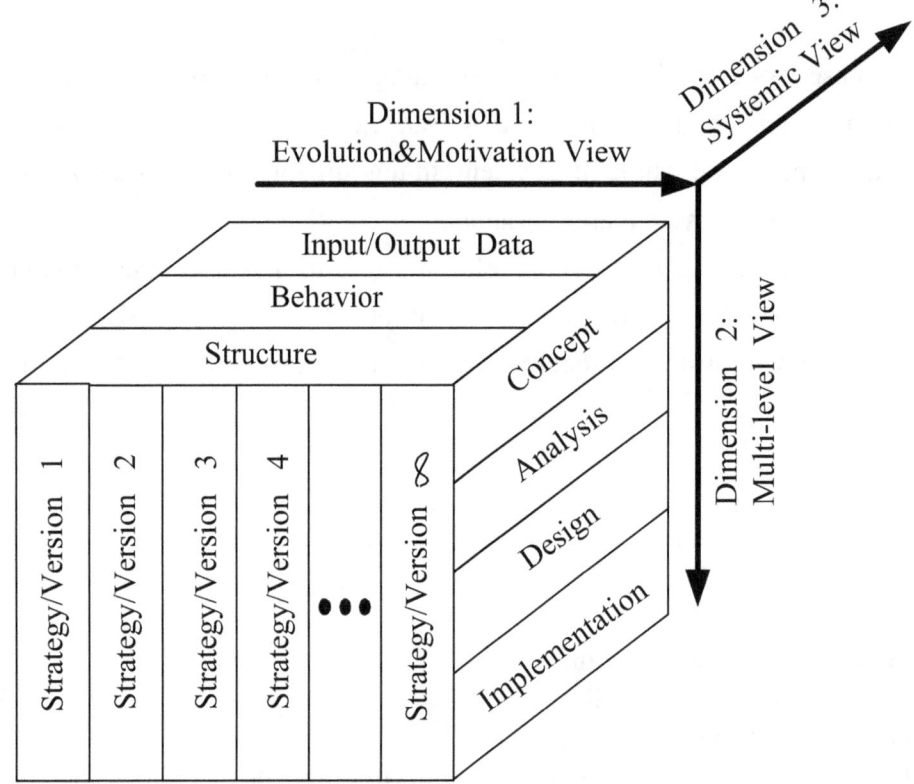

Figure 5-2   Three Significant Dimensions for Systems Definition 2.0
to Define a System

Dimension 1 stands for the evolution&motivation view which contains the strategy/version 1, strategy/version 2, strategy/version 3, strategy/version 4,…, and strategy/version ∞ views. Dimension 2 stands for the multi-level (hierarchical) view which contains the concept, analysis, design, and implementation views. Dimension 3 stands for the systemic view which contains the structure, behavior, and input/output data views.

5-1 Systemic View

In the SBC architecture, systemic view contains the structure, behavior, and input/output data views as shown in Figure 5-3.

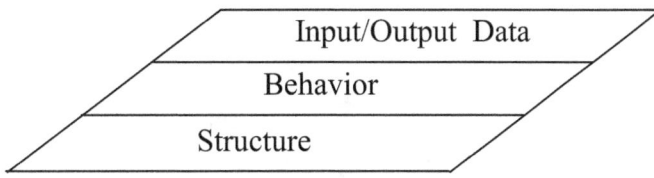

Figure 5-3   Systemic View

According to the SBC approach, the collection of all interaction flow diagrams defines the integration of the structure, behavior, and input/output data views. Adding these ideas to Figure 5-3, we then get the SBC systemic view as shown in Figure 5-4.

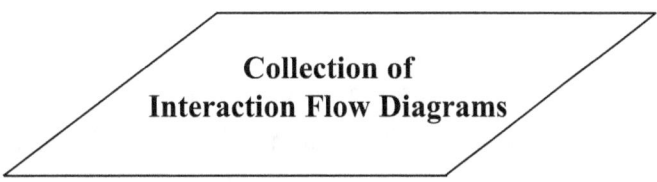

Figure 5-4   SBC Systemic View

5-2 Multi-Level View

In the SBC architecture, multi-level (hierarchical) view contains the concept, analysis, design, and implementation views as shown in Figure 5-5.

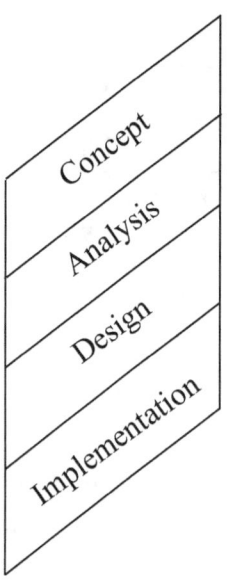

Figure 5-5   Multi-Level (Hierarchical) View

Concept view corresponds to an executive summary for an administrator who wants an estimate of the scope of the system, what it would cost, and how it would relate to the general environment in which it will operate. Analysis view corresponds to a summary for an analyzer who works on the analysis of a system. Analysis view is one level down structural decomposition (with observation congruence verification) of the concept view [Chao15c, Chao15g, Chao15h]. Design view describes what a designer has accomplished for his task. Design view is one level down structural decomposition (with observation congruence verification) of the analysis view [Chao15c, Chao15g, Chao15h]. Implementation view shows what an implementer has done for his work. Implementation view is one level down structural decomposition (with observation congruence verification) of the design view [Chao15c, Chao15g, Chao15h].

5-3 Evolution&Motivation View

A system, not matter it is physical or virtual, will always change from time to time. A system evolves when it changes. Evolution of a system is shown in Figure 5-6.

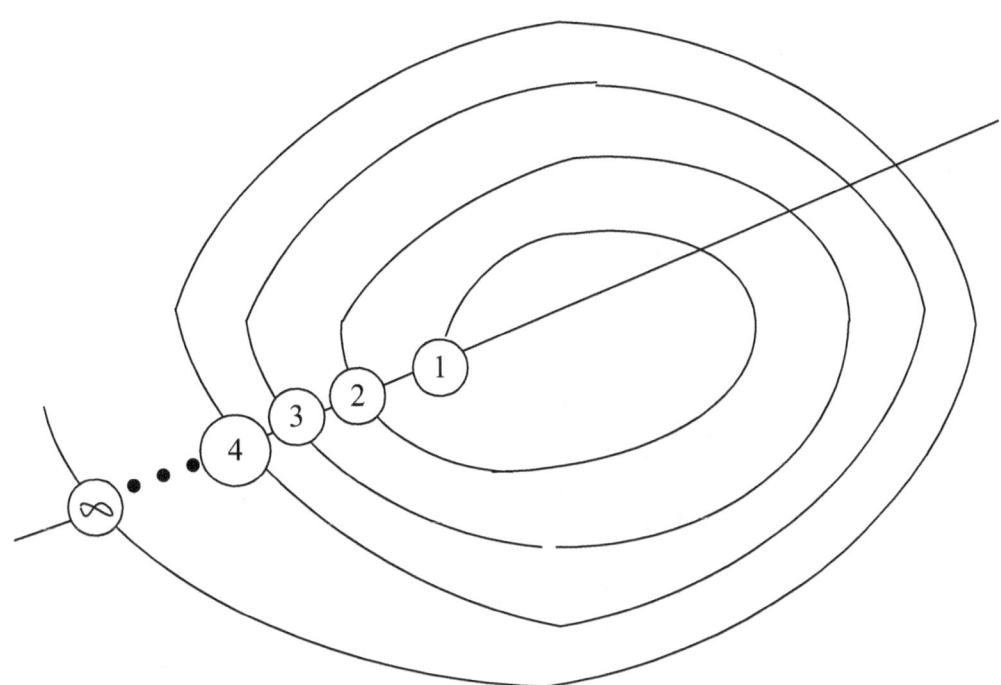

① : Systems Definition Version 1

② : Systems Definition Version 2

③ : Systems Definition Version 3

④ : Systems Definition Version 4

● ● ●

∞ : Systems Definition Version ∞

Figure 5-6   Evolution of a System

Each time when a system changes or evolves, we shall get a new version of its systems definition. In the above figure, *version 1* stands for the original systems definition of a system and evolves into *version 2*, *version 3*, *version 4*,…, and *version ∞* gradually.

Evolution of a system is represented by the SBC architecture, as the evolution&motivation view shown in Figure 5-7.

Dimension 1:
Evolution&Motivation View

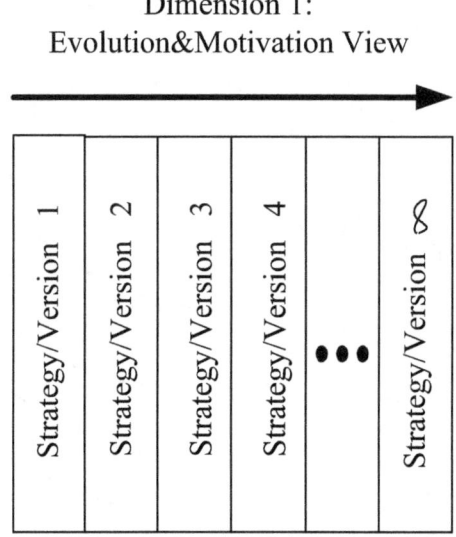

Figure 5-7   Evolution&Motivation View

In the evolution&motivation view, we see that each strategy is mapped to a version of systems definition as shown in Figure 5-8.

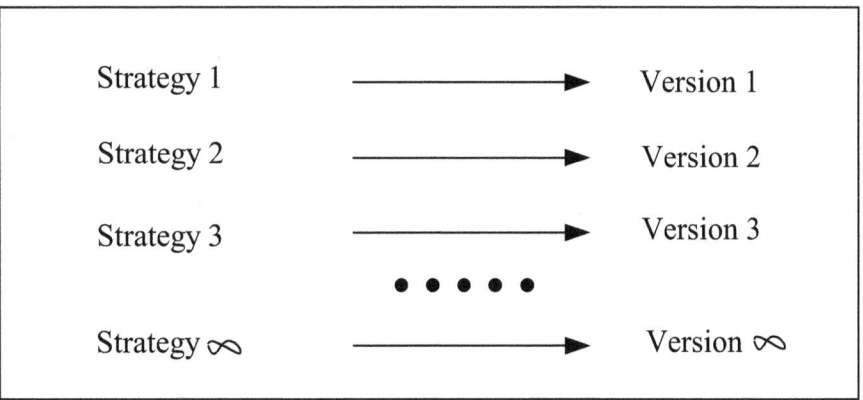

Figure 5-8   Each Strategy is Mapped to a Version of
Systems Definition

PART II: MULTI-QUEUE SBC PROCESS ALGEBRA

Chapter 6: Process Algebra as the Architecture Description Language

An architecture description is a formal description and representation of a system. A description of the systems architecture has to grasp the essence of the system and its details at the same time. In other words, an architecture description not only provides an overall picture that summarizes the whole system, but also contains enough detail that the system can be constructed and validated.

The language for architecture description is called the architecture description language (ADL) [Clem02, Clem10, Dike01, Roza05, Shaw96, Tayl09]. An ADL is a special kind of language used in describing the architecture of a system.

Process algebras are a diverse family of related approaches to the study of concurrent systems [Berg87, Hoar85, Miln89, Miln99]. Their tools are algebraic languages for the high-level description of interactions, communications, and synchronizations among independent processes. As "interaction" plays an important factor in integrating the systems structure and systems behavior for a system, it is reasonable that we use a process algebra approach as the architecture description language.

Since the architectural approach uses a coalescence model for all multiple views of a system, the foremost duty of this process algebra approach is to make the strategy/version n, strategy/version n+1, concept, analysis, design, implementation, structure, behavior, and input/output data views all integrated and coalesced within this architecture description.

6-1 What is Process Algebra?

Process algebras are a diverse family of related approaches to the study of concurrent systems. Their tools are algebraic languages for the high-level description of interactions, communications, and synchronizations among independent processes.

Process algebras also provide algebraic laws that allow process descriptions to be manipulated and analyzed, and permit formal reasoning about equivalences and observation congruence among processes.

6-2 Examples of Process Algebras

There are several leading and popular algebraic approaches to modeling concurrent systems.

Communicating Sequential Processes (CSP) was first described in a 1978 paper by C. A. R. Hoare [Hoar85].

Arthur John Robin Gorell Milner introduced the Calculus of Communicating Systems (CCS) around 1980 [Miln89].

Algebra of Communicating Processes (ACP) was initially developed by Jan Bergstra and Jan Willem Klop in 1982 [Berg87].

6-3 Multi-Queue SBC Process Algebra

Single-queue SBC process algebra (S-SBC-PA) [Chao15e, Chao15g], multi-queue SBC process algebra (M-SBC-PA) [Chao15f, Chao15h], and infinite-queue SBC process algebra (I-SBC-PA) [Chao15b, Chao15c] are the three SBC process algebras [Chao15d].

Multi-queue SBC process algebra evolved from CCS (Calculus of Communicating Systems).

CCS is a general process algebra language for the study of concurrent systems. Unlike CCS, multi-queue SBC process algebra is only applicable to systems architecture.

75

6-4 Usage of Multi-Queue SBC Process Algebra

Generally, multi-queue SBC process algebra is used to describe the three-dimensional matrix representation of multiple views as shown in Figure 6-1. That is, multi-queue SBC process algebra will describe the systemic view (dimension 3) which contains the collection of all interaction flow diagrams, and also describe the multi-level view (dimension 2) which contains the concept, analysis, design, and implementation views, and last describe the evolution&motivation view (dimension 1) which contains the strategy/version 1, strategy/version 2, strategy/version 3, strategy/version 4,…, and strategy/version ∞ views.

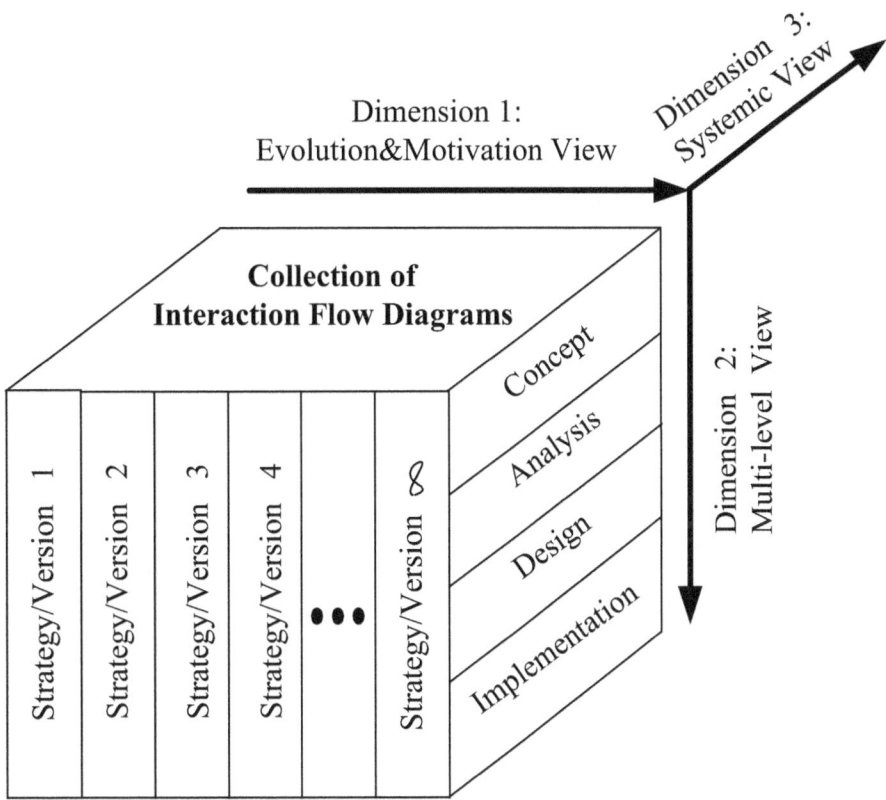

Figure 6-1   Multi-Queue SBC Process Algebra to Describe the Three-Dimensional Matrix Representation of Multiple Views

Chapter 7: Mathematics of Multi-Queue SBC Processes

To give the multi-queue SBC process a mathematical definition, we may start with a set of channels or operations whose purpose is to provide means of interaction. In addition to channels or operations, one needs a means to form new processes from old ones. The basic operators, always present in some form or other, allow sequence composition of processes or parallel composition of processes or recursive definition of a process or conditional definition of a process or renaming of a process or null process.

7-1 Interaction

An interaction represents an indivisible and instantaneous handshake or communication between two agents [Hoar85, Miln89, Miln99]. In the channel-based approach [Chao15a], the caller agent (either external environment's actor or component) interacts with the callee agent (component) through the channel interaction. In the operation-based approach [Chao15a], the caller agent (either external environment's actor or component) interacts with the callee agent (component) through the operation call or operation return interaction.

7-1-1 Channel-Based Value-Passing Interactions

Channels are a model for agent communication. An agent may provide many channels, as shown in Figure 7-1.

Figure 7-1 A Channel Model

A channel may contain several input parameters (e.g. i_1, i_2) and output parameters (e.g. o_1, o_2), as shown in Figure 7-2.

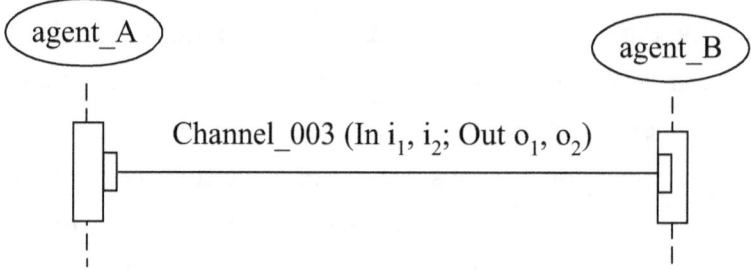

Figure 7-2 A Channel Contains Several Input/Output Parameters

A channel formula is used to completely describe a channel. A channel formula includes a) channel name, b) input parameters (e.g. i_1, i_2, ..., i_m), and c) output parameters (e.g. o_1, o_2, ..., o_n), as shown in Figure 7-3.

Channel_Name (In i_1, i_2, ..., i_m; Out o_1 , o_2, ..., o_n)

Figure 7-3   Channel Formula

An interaction represents an indivisible and instantaneous communication or handshake between two agents. In the channel-based approach as shown in Figure 7-4, the caller agent (either external environment's actor or component) interacts with the callee agent (component) through the channel interaction.

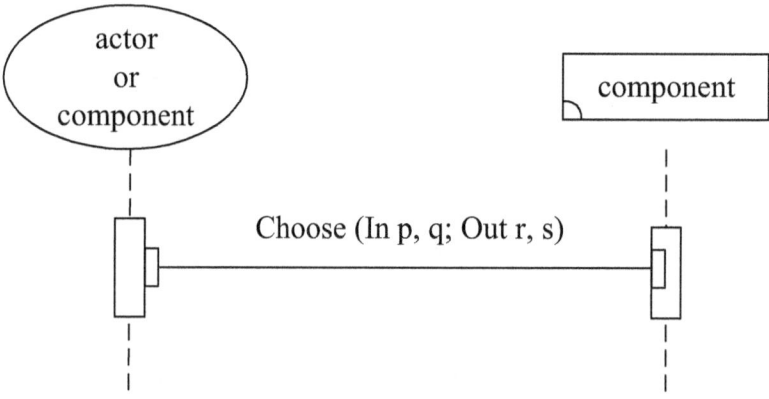

Figure 7-4   Channel-Based Value-Passing Interaction

The caller agent owns the "calling port" of the interaction. In this case, the calling port is " 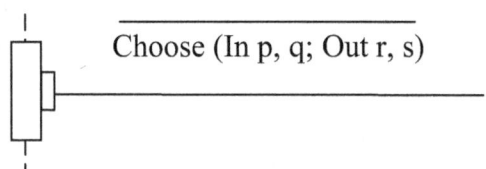 " and its conduct is to assist the caller agent to output a value to each of the "p" and "q" variables (of the "Choose" channel), and input a value from each of the "r" and "s" variables (of the "Choose" channel), as shown in Figure 7-5.

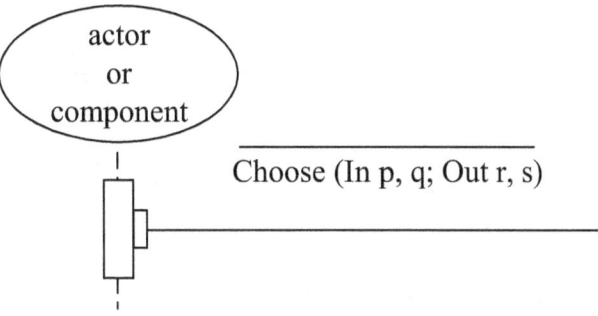

Figure 7-5   Calling Port

The caller agent together with the "calling port" is named the "calling action" as shown in Figure 7-6.

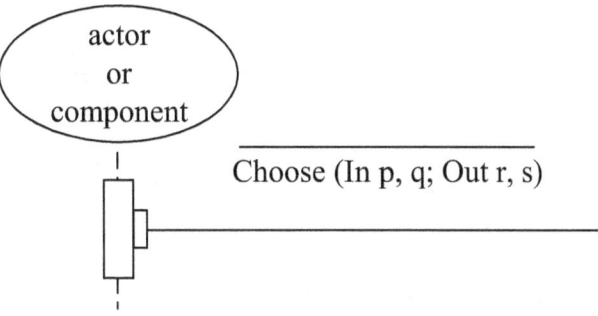

Figure 7-6   Calling Action

The callee agent owns the "called port" of the interaction. In this case, the called port is "Choose (In p, q; Out r, s)" and its conduct is to assist the callee agent to input a value from each of the "p" and "q" variables (of the "Choose" channel), and output a value to each of the "r" and "s" variables (of the "Choose" channel), as shown in Figure 7-7.

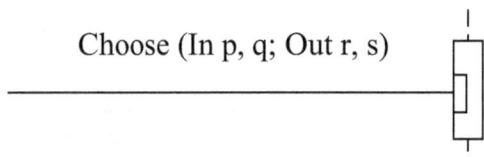

Figure 7-7   Called Port

The callee agent together with the "called port" is named the "called action" as shown in Figure 7-8.

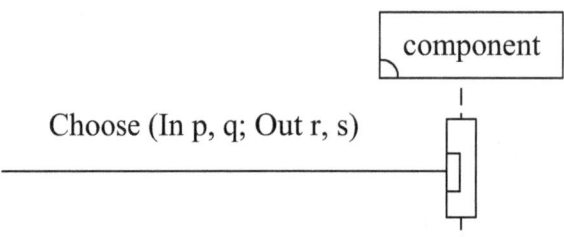

Figure 7-8   Called Action

In order to simplify the channel-based interaction diagram, we will redraw it as shown in Figure 7-9.

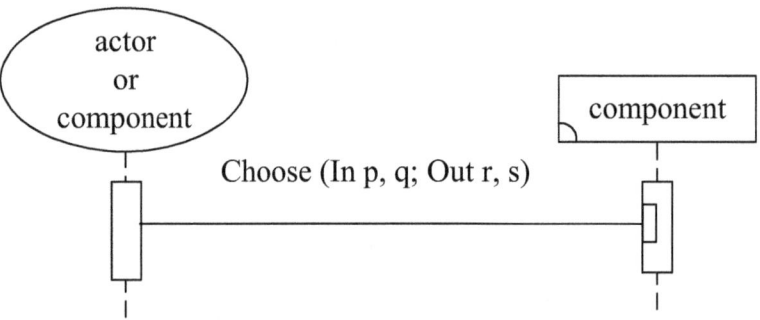

Figure 7-9   Channel-Based Interaction Diagram (I)

Or we can draw the channel-based interaction diagram as shown in Figure 7-10.

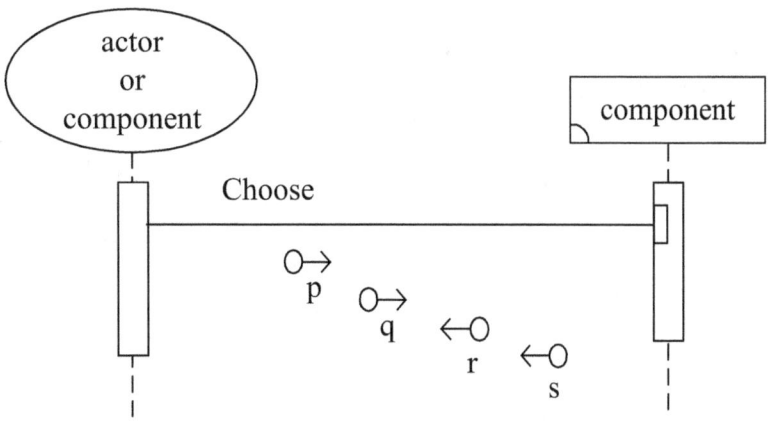

Figure 7-10   Channel-Based Interaction Diagram (II)

We use an internal interaction (i.e. λ) to represent their handshake or communication, if the caller agent and the callee agent are the same component as shown in Figure 7-11.

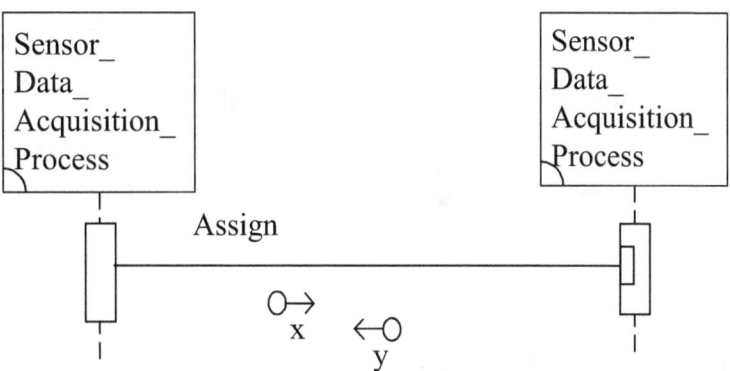

Figure 7-11   An Internal Interaction (I)

Also, we may redraw the internal interaction as shown in Figure 7-12.

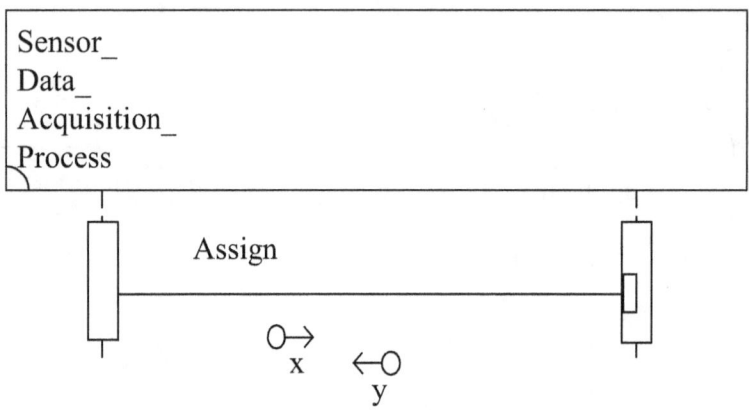

Figure 7-12   An Internal Interaction (II)

7-1-2 Operation-Based Value-Passing Interactions

An operation provided by each component represents a procedure, or method, or function of the component as shown in Figure 7-13.

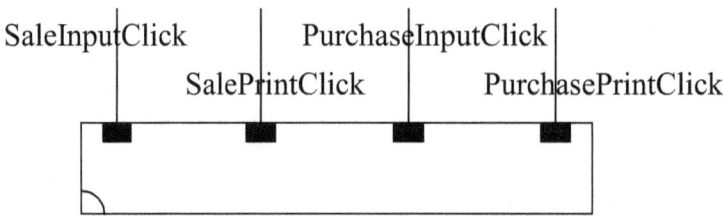

Figure 7-13   An Operation Represents a Procedure, or Method, or Function of a Component

An operation may contain several input parameters (e.g. i_1, i_2) and output parameters (e.g. o_1, o_2), as shown in Figure 7-14.

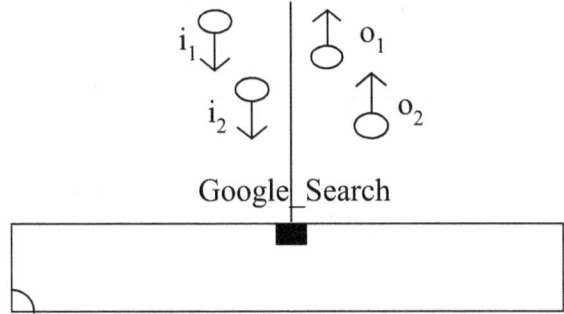

Figure 7-14   An Operation Contains Several Input/Output Parameters

An operation formula is used to completely describe an operation. An operation formula includes a) operation name, b) input parameters (e.g. i_1, i_2, ..., i_m), and c) output parameters (e.g. o_1, o_2, ..., o_n), as shown in Figure 7-15.

Operation_Name (In i_1, i_2, ..., i_m; Out o_1 , o_2, ..., o_n)

Figure 7-15   Operation Formula

An interaction represents an indivisible and instantaneous handshake or communication between two agents. In the operation-based approach as shown in Figure 7-16, the caller agent (either external environment's actor or component) communicates with the callee agent (component) through the operation call or return interaction (also named as operation call or reply message).

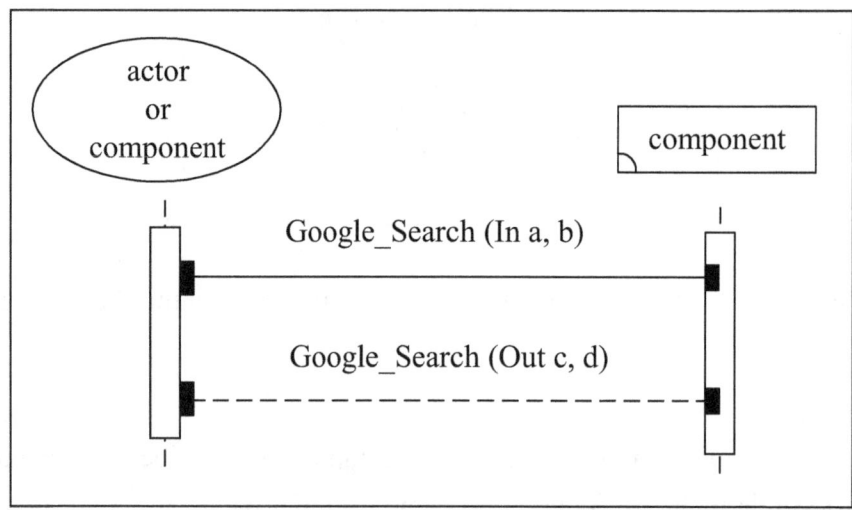

Figure 7-16   Operation-Based Value-Passing Interaction

The caller agent owns the "calling port" of the interaction. In the operation call interaction (also known as operation call message) case, the calling port is "

Google_Search (In a, b)   " and its conduct is to assist the caller agent to output a value to each of the "a" and "b" variables (of the "Google_Search" operation), as shown in Figure 7-17.

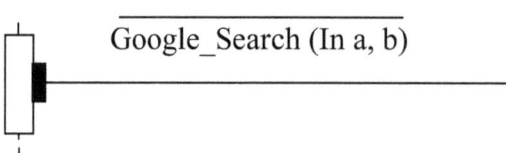

Figure 7-17   Calling Port
in the Operation Call Interaction Case

The caller agent together with the "calling port" is named the "calling action" as shown in Figure 7-18.

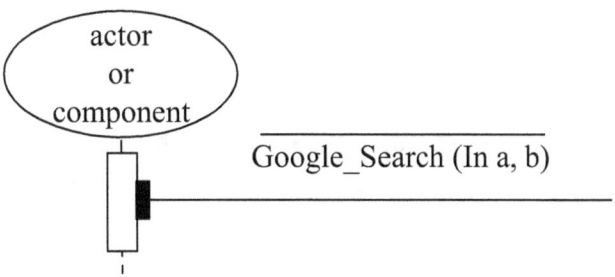

Figure 7-18   Calling Action
in the Operation Call Interaction Case

In the operation return interaction (also known as operation reply message) case, the calling port is "  " and its conduct is to assist the caller agent to input a value from each of the "c" and "d" variables (of the "Google_Search" operation), as shown in Figure 7-19.

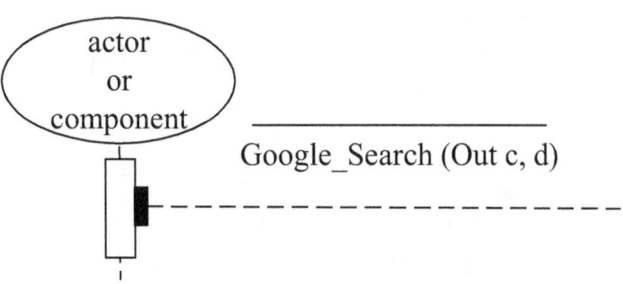

Figure 7-19   Calling Port
in the Operation Return Interaction Case

The caller agent together with the "calling port" is named the "calling action" as shown in Figure 7-20.

Figure 7-20   Calling Action
in the Operation Return Interaction Case

The callee agent owns the "called port" of the interaction. In the operation call interaction case, the called port is "Google_Search (In a, b)" and its conduct is to assist the callee agent to input a value from each of the "a" and "b" variables (of the "Google_Search" operation), as shown in Figure 7-21.

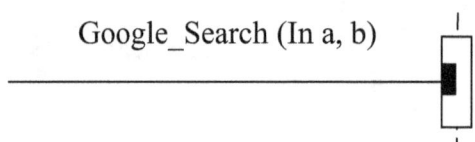

Figure 7-21   Called Port
in the Operation Call Interaction Case

The callee agent together with the "called port" is named the "called action" as shown in Figure 7-22.

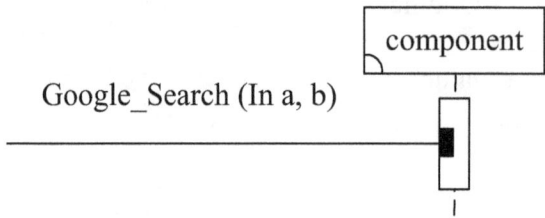

Figure 7-22   Called Action
in the Operation Call Interaction Case

In the operation return interaction case, the called port is "Google_Search (Out c, d)" and its conduct is to assist the callee agent to output a value to each of the "c" and "d" variables (of the "Google_Search" operation), as shown in Figure 7-23.

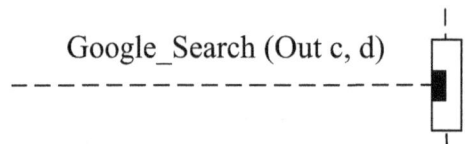

Figure 7-23   Called Port
in the Operation Return Interaction Case

The callee agent together with the "called port" is named the "called action" as shown in Figure 7-24.

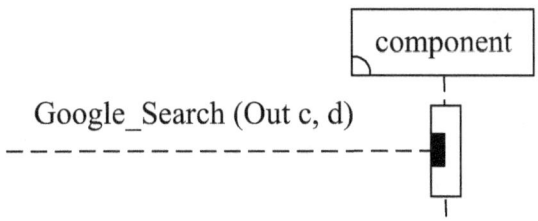

Figure 7-24   Called Action
in the Operation Return Interaction Case

In order to simplify the operation-based interaction diagram, we will redraw it as shown in Figure 7-25.

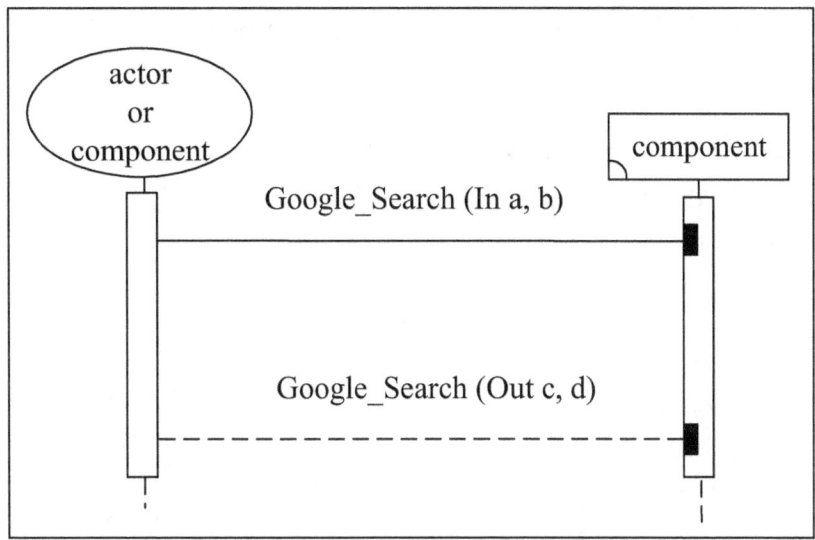

Figure 7-25   Operation-Based Interaction Diagram (I)

Or we can draw the operation-based interaction diagram as shown in Figure 7-26.

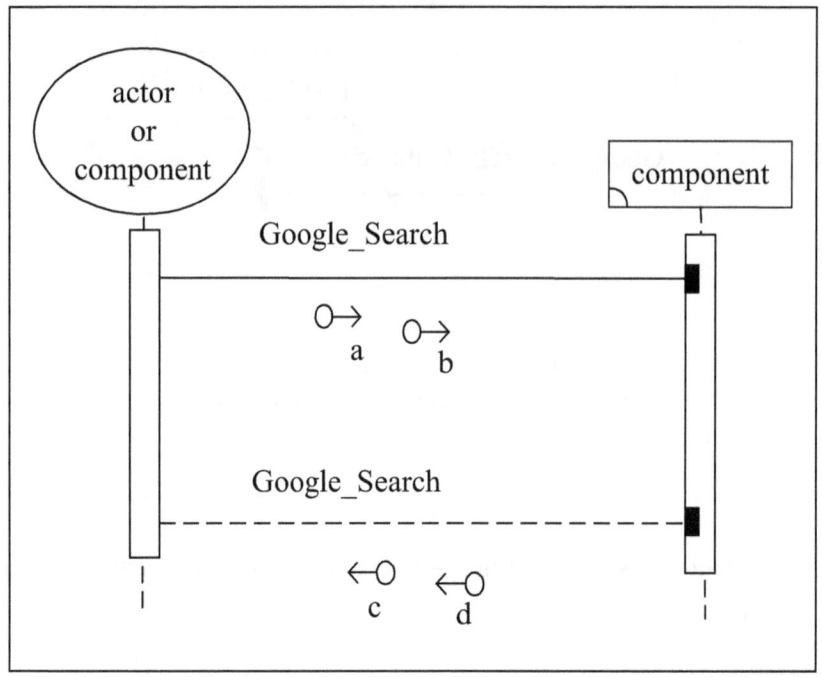

Figure 7-26   Operation-Based Interaction Diagram (II)

We use an internal interaction   (i.e. λ) to represent their handshake or communication, if the caller agent and the callee agent are the same component as shown in Figure 7-27.

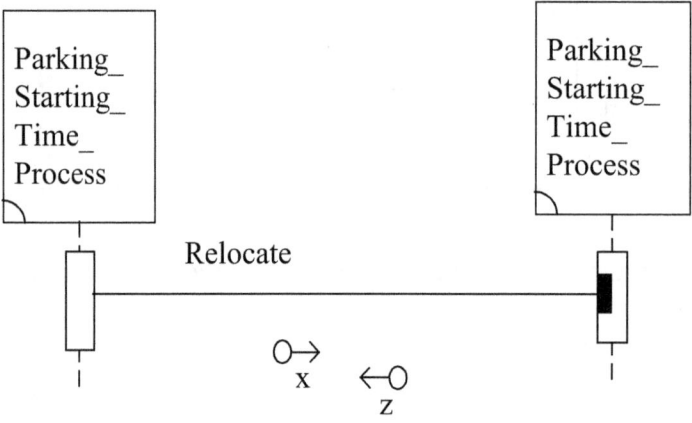

Figure 7-27   An Internal Interaction (I)

Also, we may redraw the internal interaction as shown in Figure 7-28.

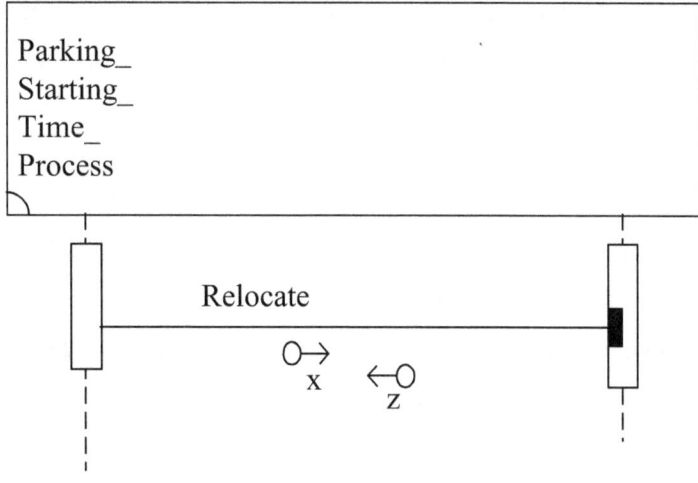

Figure 7-28   An Internal Interaction (II)

7-2 Sequentialization of Prefixes

Sometimes prefixes must be temporally ordered. For example, it might be desirable to specify algorithms such as: execute the "a" prefix first and then execute the "P" process later. Sequentialization of prefixes can be used for such purposes.

Sequentialization of prefixes, usually written as the a•P process, indicates that it will perform the "a" prefix first and continue as the "P" process.

7-3 Parallel Composition of Processes

Parallel composition of two processes P and Q, usually written $P\|Q$, is the key primitive distinguishing the process algebras from sequential models of process executions.

Parallel composition allows the executions in P and Q to proceed simultaneously and independently.

7-4 Recursive Definition of a Process

The operators presented so far describe only finite interaction and are

90

consequently insufficient for full computability, which includes non-terminating behavior. Recursion is the operation that allows finite descriptions of infinite behavior.

For example, **fix**(X=z) can be understood as abbreviating the recursive definition of an infinite behavior denoted by the "X" process variable.

7-5 Conditional Definition of a Process

A process can be defined by a one-or-more-armed conditional expression. For example, the process (**if** $cond_1$ **then** P_1)+(**if** $cond_2$ **then** P_2)...+(**if** $cond_j$ **then** P_j) will proceed as the process P_1 if the "$cond_1$" value is true, or proceed as the process P_2 if the "$cond_2$" value is true,..., or proceed as the process P_j if the "$cond_j$" value is true.

7-6 Renaming of a Process

Structural composition is accomplished by the renaming function f, or renaming combinatory [f].

For each renaming function f, the renaming combinatory [f], postfixed to a process or an interaction, has the effect of renaming the components (of the process or interaction) as dictated by f. We often write $C'_1/C_1,..., C'_n/C_n$ for the renaming function for which $f(C_i) = C'_i$ for i = 1,..., n.

7-7 Null Process

Process algebras generally also include a null process, denoted as *STOP*, which has no interaction points. It is utterly inactive and its sole purpose is to act as the inductive anchor on top of which more interesting processes can be generated.

The process "*STOP•P*" (i.e. sequence composition of processes *STOP* and *P*) equals to the process "*STOP*", as shown in Figure 7-29.

$$STOP \bullet P_1 \quad = \quad STOP$$

Figure 7-29   Characteristics of Null Process (I)

The process "$P \parallel STOP$" (i.e., parallel composition of processes P and $STOP$) equals to the process "$STOP \parallel P$" (i.e., parallel composition of processes $STOP$ and P) which equals to the process "P", as shown in Figure 7-30.

$$P \parallel STOP \quad = \quad STOP \parallel P \quad = \quad P$$

Figure 7-30   Characteristics of Null Process (II)

PART III: SYSTEMIC VIEW

Chapter 8: Multi-Queue SBC Process Algebra Language Constructs Regarding the Systemic View

In the chapter, we illustrate in detail those multi-queue SBC process algebra language constructs which make up the systemic view.

8-1 Backus-Naur Form for the Systemic View

The set of multi-queue SBC (i.e. Structure-Behavior Coalescence) process for the systemic view is defined by the following BNF grammar, as shown in Figure 8-1.

(1) <Multi-Queue_SBC_Process_of_Systemic_View> ::=
      <Parallel_of_FixIFD>

(2) <Parallel_of_FixIFD> ::=   *STOP*
  | <FixIFD> " ‖ " <Parallel_of_FixIFD>

(3) <FixIFD> ::= **fix**("<Process_Variable>"="
             <IFD><Process_Variable>")"

(4) <IFD> ::=
    <Type_1_Expression> <Zero_Or_More_Expressions>

(5) <Zero_Or_More_Expressions> ::= " ● "
  | " ● " <Type_1_Or_2_Expression> <Zero_Or_More_Expressions>

(6) <Type_1_Or_2_Expression> ::= <Type_1_Expression>
  | <Type_2_Expression>

(7) <Type_1_Expression> ::= <Type_1_Interaction>
  | <Condition> <Type_1_Interaction>
    {"+" <Condition> <Type_1_Interaction>}

(8) <Type_2_Expression> ::= <Type_2_Interaction>
  | <Condition> <Type_2_Interaction>
    {"+" <Condition> <Type_2_Interaction>}

(9) <Type_1_Interaction> ::= <Actor> <Operation_Call_Or_Return>
      <Operation_Call_Or_Return_Formula> <Component>

(10) <Type_2_Interaction> ::= <Component> <Operation_Call_Or_Return>
      <Operation_Call_Or_Return_Formula> <Component>

Figure 8-1   Backus-Naur Form for the Systemic View

8-2 Parallel of All Recursive Interaction Flow Diagrams Defines the Multi-Queue SBC Process of the Systemic View

Rule 1 describes that the parallel of all recursive interaction flow diagrams (i.e. Parallel_of_FixIFD), in which each interaction flow diagram may loop itself a countably infinite times, defines the multi-queue SBC process of the systemic view, as shown in Figure 8-2.

Rule 1
<Multi-Queue_SBC_Process_for_Systemic_View> ::=   <Parallel_of_FixIFD>

Figure 8-2   Rule 1

8-3 A NULL Process or the Parallel Composition Composing a Recursive Interaction Flow Diagram and the Parallel of All Recursive Interaction Flow Diagrams Defines the Parallel of All Recursive Interaction Flow Diagrams

Rule 2 describes that we use either a) a null process (i.e. *STOP*), or b) the parallel composition (i.e. $\parallel$) composing a recursive interaction flow diagram (i.e. FixIFD) and the parallel of all recursive interaction flow diagrams (i.e. Parallel_of_FixIFD), to define the parallel of all recursive interaction flow diagrams (i.e. Parallel_of_FixIFD), as shown in Figure 8-3.

Rule 2
<Parallel_of_FixIFD> ::=   *STOP* | <FixIFD> " $\parallel$ " <Parallel_of_FixIFD>

Figure 8-3   Rule 2

8-4 Recursion of an Interaction Flow Diagram Defines the Recursive Interaction Flow Diagram

Rule 3 describes that we use the recursion (i.e. **fix**) of an interaction flow diagram (i.e. IFD) to define a recursive interaction flow diagram (i.e. FixIFD), as shown in Figure 8-4.

Rule 3
<FixIFD> ::=   **fix**("<Process_Variable>"="   <IFD><Process_Variable>")"

Figure 8-4   Rule 3

8-5 An Interaction Flow Diagram Consists of a Type_1 Expression, Followed by Zero or More Expressions

Rule 4 describes that an interaction flow diagram (i.e. IFD) consists of a type_1 expression (i.e. Type_1_Expression) and followed by zero or more expressions (i.e. Zero_Or_More_Expressions), as shown in Figure 8-5.

Rule 4
<IFD> ::=   <Type_1_Expression> <Zero_Or_More_Expressions>

Figure 8-5   Rule 4

8-6 Zero or More Expressions either are a Sequence Composition or Consist of a Sequence Composition, Followed by a Type_1_Or_2 Expression, and Followed by Zero or More Expressions

Rule 5 describes that zero or more expressions (i.e. Zero_Or_More_Expressions) either a) are a sequence composition (i.e. ●), or b) consist of a sequence composition (i.e. ●), followed by a type_1_or_2 expression (i.e. Type_1_Or_2_Expression), and followed by zero or more expressions (i.e. Zero_Or_More_Expressions), as shown in Figure 8-6.

Rule 5
<Zero_Or_More_Expressions> ::= " ●"<Process_Variable> 	" ●" <Type_1_Or_2_Expression> <Zero_Or_More_Expressions>

Figure 8-6   Rule 5

8-7 Type_1_Or_2 Expression is either Type_1 or Type_2

Rule 6 describes that the type_1_or_2 expression (i.e. Type_1_Or_2_Expression) is either a type_1 expression (i.e. Type_1_Expression) or a type_2 expression (i.e. Type_2_Expression)., as shown in Figure 8-7.

Rule 6
<Type_1_Or_2_Expression> ::=   <Type_1_Expression> | <Type_2_Expression>

Figure 8-7   Rule 6

8-8 Type_1 Expression is either an Unconditional Type_1 Interaction or a Conditional Type_1 Interaction

Rule 7 describes that the type_1 expression (i.e. Type_1_Expression) is either an unconditional type_1 interaction (i.e. Type_1_Interaction) or a conditional type_1 interaction (i.e. one-or-more-armed conditional expression of Type_1_Interaction), as shown in Figure 8-8.

```
Rule 7

<Type_1_Expression>    ::=  <Type_1_Interaction>
 | <Condition> <Type_1_Interaction>
    {"+" <Condition> <Type_1_Interaction>}
```

Figure 8-8   Rule 7

8-9 Type_2 Expression is either an Unconditional Type_2 Interaction or a Conditional Type_2 Interaction

Rule 8 describes that the type_2 expression (i.e. Type_2_Expression) is either an unconditional type_2 interaction (i.e. Type_2_Interaction) or a conditional type_2 interaction (i.e. one-or-more-armed conditional expression of Type_2_Interaction), as shown in Figure 8-9.

```
Rule 8

<Type_2_Expression>    ::=  <Type_2_Interaction>
 | <Condition> <Type_2_Interaction>
    {"+" <Condition> <Type_2_Interaction>}
```

Figure 8-9   Rule 8

8-10 An Actor Interacting with a Component Defines the Type_1 Interaction

Rule 9 describes that an actor interacting with a component defines the type_1 interaction, as shown in Figure 8-10.

Rule 9
<Type_1_Interaction> ::= <Actor> <Operation_Call_Or_Return> <Operation_Call_Or_Return_Formula> <Component>

Figure 8-10   Rule 9

8-11 A Component Interacting with another Component Defines the Type_2 Interaction

Rule 10 describes that a component interacting with another component defines the type_2 interaction, as shown in Figure 8-11.

Rule 10
<Type_2_Interaction> ::= <Component> <Operation_Call_Or_Return> <Operation_Call_Or_Return_Formula> <Component>

Figure 8-11   Rule 10

Chapter 9: Multi-Queue SBC Process Algebra Transitional Semantics Regarding the Systemic View

In the chapter, we illustrate in detail those multi-queue SBC process algebra transitional semantics which regards the systemic view.

9-1 Transitional Semantics for the Systemic View

As shown in Figure 9-1, we assume an infinite set Δ of type_1_or_2 interactions, and use a_1, a_2...to range over Δ. Further, we let X be the set of process variables, and use X_1, X_2...to range over X. We let Φ be the set of process Constants, and use A_1, A_2...to range over Φ. We let Π be the set of processes, and use P_1, Q_1...to range over Π. We let Ψ be the set of process expressions, and use E_1, E_2...to range over Ψ.

Entity set	Entity name	Type of entity
Δ	a_1, a_2...	type_1_or_2 interactions
X	X_1, X_2...	process variables
Φ	A_1, A_2...	process Constants
Π	P_1, Q_1...	processes
Ψ	E_1, E_2...	process expressions

Figure 9-1   Entities

In giving meaning to the multi-queue SBC process algebra for the systemic view, we shall use the following labelled transition system (LTS) [Miln89, Miln99]

$$(\Psi, \Delta, \rightarrow)$$

which consists of a set Ψ of process expressions, a set Δ of "type_1_or_2 or internal interactions", and a transition relation $\rightarrow \subseteq \Psi \times \Delta \times \Psi$ where $(E_i, a, E_j) \in \rightarrow$ is denoted by $E_i \xrightarrow{a} E_j$.

The semantics for Ψ consists in the transition rules of each transition relation $\rightarrow$ over $\Psi \times \Delta \times \Psi$. These transition rules will follow the structure of expressions.

As shown in Figure 9-2, we give the complete set of transition rules; the names Prefix, Parallel, Recursion, and Constant indicate that the rules are associated respectively with Prefix, Parallel Composition, and Recursion and with Constants.

Prefix
$$\frac{}{a \bullet E \xrightarrow{a} E}$$

Parallel$_1$
$$\frac{E \xrightarrow{a} E'}{E \parallel F \xrightarrow{a} E' \parallel F}$$

Parallel$_2$
$$\frac{F \xrightarrow{a} F'}{E \parallel F \xrightarrow{a} E \parallel F'}$$

Recursion
$$\frac{\mathbf{fix}(X=z\{\mathbf{fix}(X=z)/X\}) \xrightarrow{a} E'}{\mathbf{fix}(X=z) \xrightarrow{a} E'}$$

Constant
$$\frac{P \xrightarrow{a} P'}{A \xrightarrow{a} P'} \quad (A \overset{\text{def}}{=\joinrel=} P)$$

Figure 9-2   Transition Rules for the Systemic View

9-2 Rule of Prefix

The rule for Prefix, shown in Figure 9-3, can be read as follows: Under any circumstances, we always infer $a \bullet E \xrightarrow{a} E$. That is, an expression, with an interaction prefixed to it, will use this interaction to accomplish the transition.

$$\frac{\rule{3cm}{0.4pt}}{a \bullet E \xrightarrow{a} E}$$

Figure 9-3   Rule of Prefix

9-3 Rules of Parallel Composition

There are two transition rules for parallel composition. Rule Parallel$_1$, as shown in Figure 9-4, indicates that from $E \xrightarrow{a} E'$ we shall infer $E \parallel F \xrightarrow{a} E' \parallel F$.

$$\frac{E \xrightarrow{a} E'}{E \parallel F \xrightarrow{a} E' \parallel F}$$

Figure 9-4   Rule Parallel$_1$

Rule Parallel$_2$, as shown in Figure 9-5, indicates that from $F \xrightarrow{a} F''$ we shall infer $E \| F \xrightarrow{a} E \| F''$.

$$\frac{F \xrightarrow{a} F'}{E \| F \xrightarrow{a} E \| F'}$$

Figure 9-5   Rule Parallel$_2$

9-4 Rule of Recursion

The rule for Recursion, shown in Figure 9-6, can be read as follows: This says that any interaction which may be inferred for the **fix** expression 'unwound' once (by substituting itself for its bound variable) may be inferred for the **fix** expression itself.

$$\frac{\mathbf{fix}(X=z\{\mathbf{fix}(X=z)/X\}) \xrightarrow{a} E'}{\mathbf{fix}(X=z) \xrightarrow{a} E'}$$

Figure 9-6   Rule of Recursion

9-5 Rule of Constants

The rule for Constants, shown in Figure 9-7, can be read as follows: the rule of Constants asserts that each Constant has the same transitions as its defining expression.

$$\frac{P \xrightarrow{a} P'}{A \xrightarrow{a} P'} \quad (A \stackrel{\text{def}}{=} P)$$

Figure 9-7   Rule of Constants

Chapter 10: Kurdi University as an Example

The collection of all interaction flow diagrams defines the SBC systemic view of *Kurdi University* as shown in Figure 10-1.

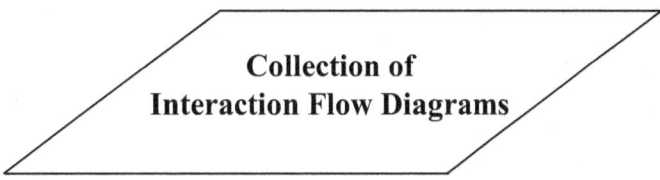

Collection of
Interaction Flow Diagrams

Figure 10-1   SBC Systemic View of Kurdi University

10-1 Overall Behavior of the Systemic View of Kurdi University

The overall behavior of the systemic view of *Kurdi University* includes two behaviors: *Study_Calculus_Course* and *Study_Algebra_Course* as shown in Figure 10-2. Each of them is described by an individual IFD.

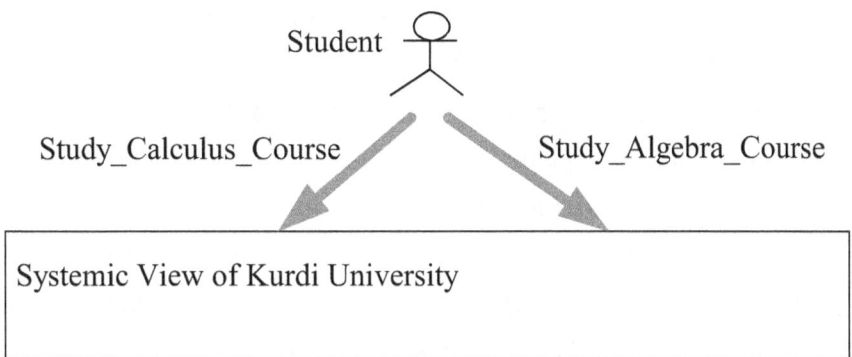

Figure 10-2   Overall Behavior of the Systemic View of Kurdi University

An IFD of the *Study_Calculus_Course* behavior is shown in Figure 10-3. First, actor *Student* interacts with the *University_President* component through the *University_Teach_Calculus* operation call interaction. Next, component *University_President* interacts with the *Science_Dean* component through the *College_Teach_Calculus* operation call interaction. Continuingly, component *Science_Dean* interacts with the *Mathematics_Chairman* component through the

Department_Teach_Calculus operation call interaction. Finally, component *Mathematics_Chairman* interacts with the *Calculus_Lecturer* component through the *Lecturer_Teach_Calculus* operation call interaction.

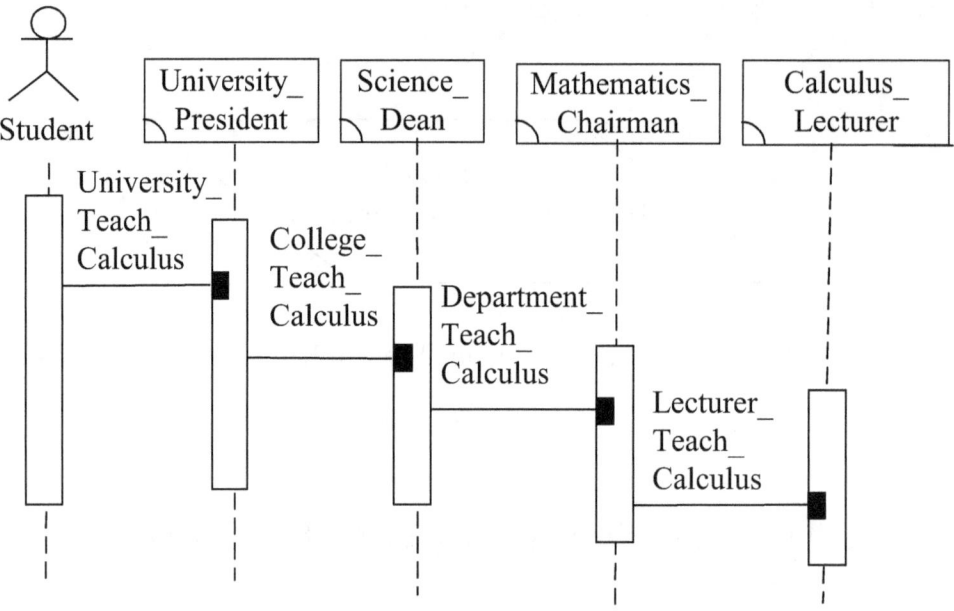

Figure 10-3   IFD of the *Study_Calculus_Course* Behavior

An IFD of the *Study_Algebra_Course* behavior is shown in Figure 10-4. First, actor *Student* interacts with the *University_President* component through the *University_Teach_Algebra* operation call interaction. Next, component *University_President* interacts with the *Science_Dean* component through the *College_Teach_Algebra* operation call interaction. Continuingly, component *Science_Dean* interacts with the *Mathematics_Chairman* component through the *Department_Teach_Algebra* operation call interaction. Finally, component *Mathematics_Chairman* interacts with the *Algebra_Lecturer* component through the *Lecturer_Teach_Algebra* operation call interaction.

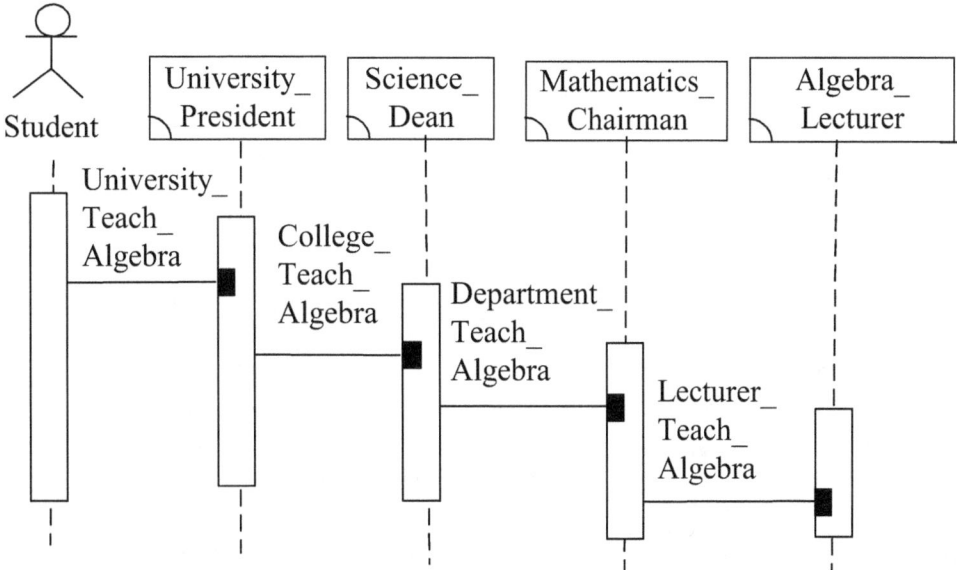

Figure 10-4 IFD of the *Study_Algebra_Course* Behavior

10-2 Backus-Naur Form for the Systemic View of Kurdi University

We draw the multi-queue SBC process algebra Backus-Naur Form tree of the systemic view of *Kurdi University* as shown in Figure 10-5.

110

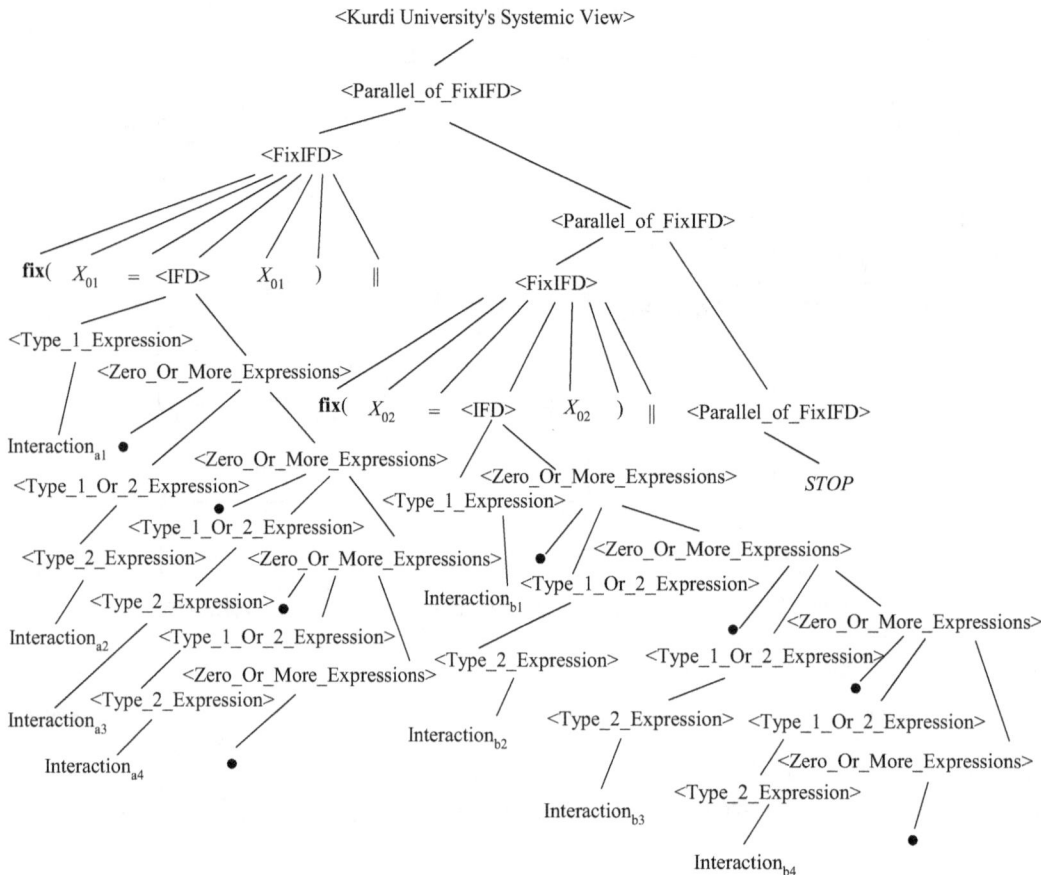

Figure 10-5   M-SBC-PA Backus-Naur Form Tree of the Systemic View of Kurdi University

10-3 Interactions of the Systemic View of Kurdi University

Interaction$_{a1}$ stands for the 1st interaction of the ath interaction flow diagram of the systemic view of *Kurdi University*, as shown in Figure 10-6. Interaction$_{a1}$ is a type_1 interaction which describes the *Student* actor interacts with the *University_President* component.

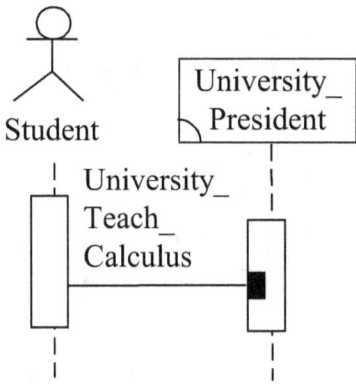

Figure 10-6   Interaction$_{a1}$

Interaction$_{a2}$ stands for the 2nd interaction of the ath interaction flow diagram of the systemic view of *Kurdi University*, as shown in Figure 10-7. Interaction$_{a2}$ is a type_2 interaction which describes the *University_President* component interacts with the *Science_Dean* component.

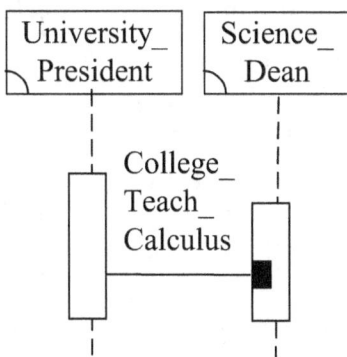

Figure 10-7   Interaction$_{a2}$

Interaction$_{a3}$ stands for the 3rd interaction of the ath interaction flow diagram of the systemic view of *Kurdi University*, as shown in Figure 10-8. Interaction$_{a3}$ is a type_2 interaction which describes the *Science_Dean* component interacts with the *Mathematics_Chairman* component.

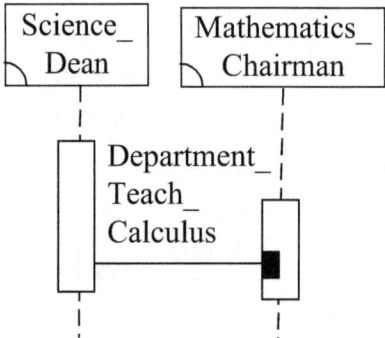

Figure 10-8   Interaction$_{a3}$

Interaction$_{a4}$ stands for the 4th interaction of the ath interaction flow diagram of the systemic view of *Kurdi University*, as shown in Figure 10-9. Interaction$_{a4}$ is a type_2 interaction which describes the *Mathematics_Chairman* component interacts with the *Calculus_Lecturer* component.

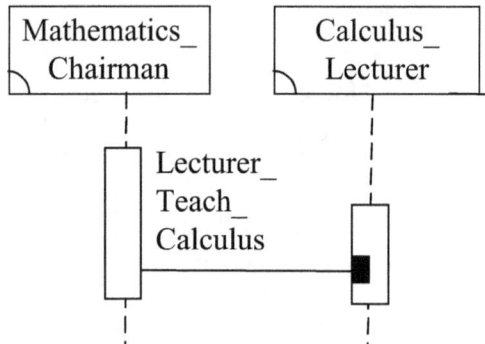

Figure 10-9   Interaction$_{a4}$

Interaction$_{b1}$ stands for the 1st interaction of the bth interaction flow diagram of the systemic view of *Kurdi University*, as shown in Figure 10-10. Interaction$_{b1}$ is a type_1 interaction which describes the *Student* actor interacts with the *University_President* component.

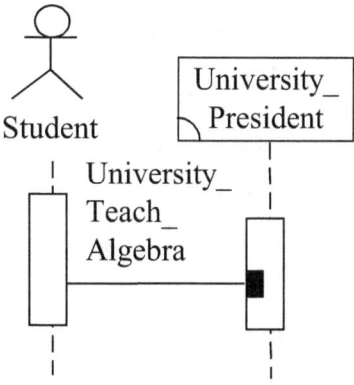

Figure 10-10   Interaction$_{b1}$

Interaction$_{b2}$ stands for the 2nd interaction of the bth interaction flow diagram of the systemic view of *Kurdi University*, as shown in Figure 10-11. Interaction$_{b2}$ is a type_2 interaction which describes the *University_President* component interacts with the *Science_Dean* component.

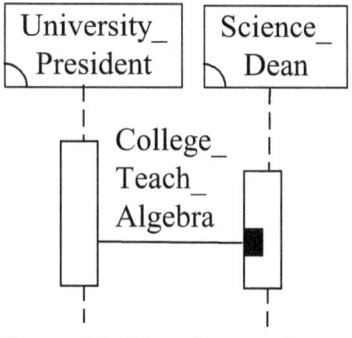

Figure 10-11   Interaction$_{b2}$

Interaction$_{b3}$ stands for the 3rd interaction of the bth interaction flow diagram of the systemic view of *Kurdi University*, as shown in Figure 10-12. Interaction$_{b3}$ is a type_2 interaction which describes the *Science_Dean* component interacts with the *Mathematics_Chairman* component.

114

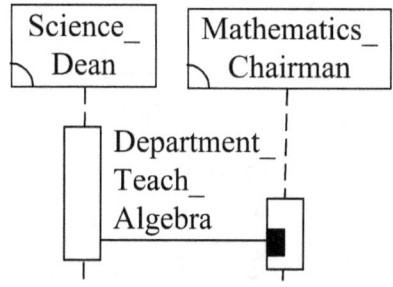

Figure 10-12   Interaction$_{b3}$

Interaction$_{b4}$ stands for the 4th interaction of the bth interaction flow diagram of the systemic view of *Kurdi University*, as shown in Figure 10-13. Interaction$_{b4}$ is a type_2 interaction which describes the *Mathematics_Chairman* component interacts with the *Algebra_Lecturer* component.

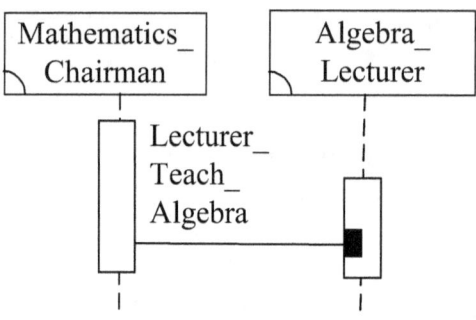

Figure 10-13   Interaction$_{b4}$

10-4 Interaction Flow Diagrams of the Systemic View of Kurdi University

FixIFD$_a$ describes the recursion of the ath interaction flow diagram, i.e. *Study_Calculus_Course* behavior, of the systemic view of *Kurdi University*. FixIFD$_a$ is syntactically represented as "**fix**(X_{01} = Interaction$_{a1}$•Interaction$_{a2}$•Interaction$_{a3}$•Interaction$_{a4}$•X_{01})", as shown in Figure 10-14.

$$\text{FixIFD}_a \overset{\text{def}}{=\!=}$$

$$\textbf{fix}(X_{01} = \text{Interaction}_{a1} \bullet \text{Interaction}_{a2} \bullet \text{Interaction}_{a3} \bullet \text{Interaction}_{a4} \bullet X_{01})$$

Figure 10-14   FixIFD$_a$

FixIFD$_b$ describes the recursion of the bth interaction flow diagram, i.e. *Study_Algebra_Course* behavior, of the systemic view of *Kurdi University*. FixIFD$_b$ is syntactically represented as "$\textbf{fix}(X_{02} = \text{Interaction}_{b1} \bullet \text{Interaction}_{b2} \bullet \text{Interaction}_{b3} \bullet \text{Interaction}_{b4} \bullet X_{02})$", as shown in Figure 10-15.

$$\text{FixIFD}_b \overset{\text{def}}{=\!=}$$

$$\textbf{fix}(X_{02} = \text{Interaction}_{b1} \bullet \text{Interaction}_{b2} \bullet \text{Interaction}_{b3} \bullet \text{Interaction}_{b4} \bullet X_{02})$$

Figure 10-15   FixIFD$_b$

10-5 Multi-Queue SBC Process of the Systemic View of Kurdi University

Multi-queue SBC process of the systemic view of *Kurdi University* is syntactically represented as FixIFD$_a$‖FixIFD$_b$ which equals to "$\textbf{fix}(X_{01} = \text{Interaction}_{a1} \bullet \text{Interaction}_{a2} \bullet \text{Interaction}_{a3} \bullet \text{Interaction}_{a4} \bullet X_{01})$ ‖ $\textbf{fix}(X_{02} = \text{Interaction}_{b1} \bullet \text{Interaction}_{b2} \bullet \text{Interaction}_{b3} \bullet \text{Interaction}_{b4} \bullet X_{02})$", as shown in Figure 10-16.

Kurdi University's Systemic View $\overset{\text{def}}{=\!=}$

$$\textbf{fix}(X_{01} = \text{Interaction}_{a1} \bullet \text{Interaction}_{a2} \bullet \text{Interaction}_{a3} \bullet \text{Interaction}_{a4} \bullet X_{01}) \ ‖$$
$$\textbf{fix}(X_{02} = \text{Interaction}_{b1} \bullet \text{Interaction}_{b2} \bullet \text{Interaction}_{b3} \bullet \text{Interaction}_{b4} \bullet X_{02})$$

Figure 10-16 Kurdi University's Systemic View

116

PART IV: MULTI-LEVEL VIEW

Chapter 11: Relationships among Different Level Views

In the SBC architecture, multi-level (hierarchical) view contains the concept, analysis, design, and implementation views as shown in Figure 11-1.

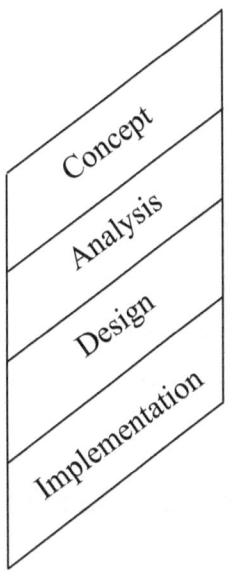

Figure 11-1   Multi-Level View

In the SBC multi-level view of systems architecture, concept view is one level up structural composition (with observation congruence verification) of the analysis view; analysis view is one level up structural composition (with observation congruence verification) of the design view; design view is one level up structural composition (with observation congruence verification) of the implementation view.

11-1 Concept View is One Level up Structural Composition of the Analysis View

To demonstrate that the concept view is one level up structural composition (with observation congruence verification) of the analysis view, we need to go through the following four steps:

(A) Get the multi-queue SBC process of the concept view.

(B) Get the multi-queue SBC process of the analysis view.

(C) Get the multi-queue SBC process of the structural composition of the analysis view.

(D) Verify that there is observation congruence of "the concept view" and "the structural composition of the analysis view".

11-2 Analysis View is One Level up Structural Composition of the Design View

To demonstrate that the analysis view is one level up structural composition (with observation congruence verification) of the design view, we need to go through the following four steps:

(A) Get the multi-queue SBC process of the analysis view.

(B) Get the multi-queue SBC process of the design view.

(C) Get the multi-queue SBC process of the structural composition of the design view.

(D) Verify that there is observation congruence of "the analysis view" and "the structural composition of the design view".

11-3 Design View is One Level up Structural Composition of the Implementation View

To demonstrate that the design view is one level up structural composition (with observation congruence verification) of the implementation view, we need to go through the following four steps:

(A) Get the multi-queue SBC process of the design view.

(B) Get the multi-queue SBC process of the implementation view.

(C) Get the multi-queue SBC process of the structural composition of the implementation view.

(D) Verify that there is observation congruence of "the design view" and "the structural composition of the implementation view".

Chapter 12: Multi-Queue SBC Process Algebra Language Constructs Regarding the Multi-Level View

In the chapter, we illustrate in detail those multi-queue SBC process algebra language constructs which make up the multi-level view.

12-1 Backus-Naur Form for (the Structural Composition of) the Systemic View of Each Level

The set of multi-queue SBC (i.e. Structure-Behavior Coalescence) processes for (the structural composition of) the systemic view of each level (i.e. Concept, Analysis, Design, Implementation) is defined by the following BNF grammar, as shown in Figure 12-1.

(1) <Structural Composition of Multi-Queue_SBC_Process_of_Each_Level's_Systemic_View> ::=
        <Multi-Queue_SBC_Process_of_Each_Level's_Systemic_View>[*f*]
(2) <Multi-Queue_SBC_Process_of_Each_Level's_Systemic_View> ::=
        <Parallel_of_FixIFD>
(3) <Parallel_of_FixIFD> ::=   *STOP*
  | <FixIFD> " ‖ " <Parallel_of_FixIFD>
(4) <FixIFD> ::= **fix**("<Process_Variable>"="
                    <IFD><Process_Variable>")"
(5) <IFD> ::=
    <Type_1_Expression> <Zero_Or_More_Expressions>
(6) <Zero_Or_More_Expressions> ::= " ● "
  | " ● " <Type_1_Or_2_Expression> <Zero_Or_More_Expressions>
(7) <Type_1_Or_2_Expression> ::= <Type_1_Expression>
  | <Type_2_Expression>
(8) <Type_1_Expression> ::= <Type_1_Interaction>
  | <Condition> <Type_1_Interaction>
    {"+" <Condition> <Type_1_Interaction>}
(9) <Type_2_Expression> ::= <Type_2_Interaction>
  | <Condition> <Type_2_Interaction>
    {"+" <Condition> <Type_2_Interaction>}
(10) <Type_1_Interaction> ::= <Actor> <Operation_Call_Or_Return>
           <Operation_Call_Or_Return_Formula> <Component>
(11) <Type_2_Interaction> ::= <Component> <Operation_Call_Or_Return>
           <Operation_Call_Or_Return_Formula> <Component>

Figure 12-1   Backus-Naur Form for the Systemic View of Each Level

12-2 Renaming the Multi-Queue SBC Process of Each Level's Systemic View Defines the Structural Composition of Multi-Queue SBC Process of the Systemic View of Each Level

Rule 1 describes that applying the renaming function (i.e. [f]) to the multi-queue SBC process of each level's (i.e. Concept, Analysis, Design, Implementation) systemic view defines the structural composition of multi-queue SBC process of the systemic view of each level (i.e. Concept, Analysis, Design, Implementation), as shown in Figure 12-2.

Rule 1
<Structural Composition of Multi-Queue_SBC_Process_of_Each_Level's_Systemic_View> ::=   <Multi-Queue_SBC_Process_of_Each_Level's_Systemic_View> [f]

Figure 12-2   Rule 1

For each renaming function f, the renaming combinatory [f], postfixed to a process or an interaction, has the effect of renaming the components of the process or interaction as dictated by f. We often write $C'_1/C_1,\ldots, C'_n/C_n$ for the renaming function for which $f(C_i) = C'_i$ for i = 1,…, n.

After renaming, an interaction may be internalized to be an internal interaction (i.e. λ) if it describes that a component interacts with itself as shown in Figure 12-3.

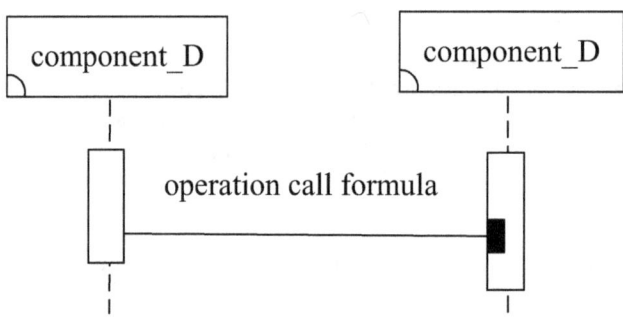

Figure 12-3   A component Interacts with Itself

12-3 Parallel of All Recursive Interaction Flow Diagrams Defines the Multi-Queue SBC Process of the Systemic View of Each Level

Rule 2 describes that the parallel of all recursive interaction flow diagrams (i.e. Parallel_of_FixIFD), in which each interaction flow diagram may loop itself a countably infinite times, defines the multi-queue SBC process of the systemic view of each level (i.e. Concept, Analysis, Design, Implementation), as shown in Figure 12-4.

Rule 2
<Multi-Queue_SBC_Process_of_Each_Level's_Systemic_View> ::=   <Parallel_of_FixIFD>

Figure 12-4   Rule 2

12-4 A NULL Process or the Parallel Composition Composing a Recursive Interaction Flow Diagram and the Parallel of All Recursive Interaction Flow Diagrams Defines the Parallel of All Recursive Interaction Flow Diagrams

Rule 3 describes that we use either a) a null process (i.e. *STOP*), or b) the parallel composition (i.e. $\parallel$) composing a recursive interaction flow diagram (i.e. FixIFD) and the parallel of all recursive interaction flow diagrams (i.e. Parallel_of_FixIFD), to define the parallel of all recursive interaction flow diagrams (i.e. Parallel_of_FixIFD), as shown in Figure 12-5.

Rule 3
<Parallel_of_FixIFD> ::= *STOP* | <FixIFD> " $\parallel$ " <Parallel_of_FixIFD>

Figure 12-5   Rule 3

12-5 Recursion of an Interaction Flow Diagram Defines the Recursive Interaction Flow Diagram

Rule 4 describes that we use the recursion (i.e. **fix**) of an interaction flow diagram (i.e. IFD) to define a recursive interaction flow diagram (i.e. FixIFD), as shown in Figure 12-6.

Rule 4
<FixIFD> ::= "**fix**("<Process_Variable>"="   <IFD><Process_Variable>")"

Figure 12-6   Rule 4

12-6 An Interaction Flow Diagram Consists of a Type_1 Expression, Followed by Zero or More Expressions

Rule 5 describes that an interaction flow diagram (i.e. IFD) consists of a type_1 expression (i.e. Type_1_Expression) and followed by zero or more expressions (i.e. Zero_Or_More_Expressions), as shown in Figure 12-7.

Rule 5
<IFD> ::=   <Type_1_Expression> <Zero_Or_More_Expressions>

Figure 12-7   Rule 5

12-7 Zero or More Expressions either are a Sequence Composition or Consist of a Sequence Composition, Followed by a Type_1_Or_2 Expression, and Followed by Zero or More Expressions

Rule 6 describes that zero or more expressions (i.e. Zero_Or_More_Expressions) either a) are a sequence composition (i.e. ●), or b) consist of a sequence composition (i.e. ●), followed by a type_1_or_2 expression (i.e. Type_1_Or_2_Expression), and followed by zero or more expressions (i.e. Zero_Or_More_Expressions), as shown in Figure 12-8.

Rule 6
<Zero_Or_More_Expressions> ::= "● "<Process_Variable> | "● " <Type_1_Or_2_Expression> <Zero_Or_More_Expressions>

Figure 12-8   Rule 6

12-8 Type_1_Or_2 Expression is either Type_1 or Type_2

Rule 7 describes that the type_1_or_2 expression (i.e. Type_1_Or_2_Expression) is either a type_1 expression (i.e. Type_1_Expression) or a type_2 expression (i.e. Type_2_Expression), as shown in Figure 12-9.

```
Rule 7

<Type_1_Or_2_Expression>  ::=
        <Type_1_Expression>
    |   <Type_2_Expression>
```

Figure 12-9   Rule 7

12-9 Type_2 Expression is either a Type_2 Interaction or a Conditional Type_2 Interaction

Rule 8 describes that the type_2 expression (i.e. Type_2_Expression) is either a type_2 interaction (i.e. Type_2_Interaction) or a conditional type_2 interaction (i.e. one-or-more-armed conditional expression of Type_2_Interaction), as shown in Figure 12-10.

```
Rule 8

<Type_1_Expression>    ::=  <Type_1_Interaction>
  |  <Condition> <Type_1_Interaction>
     {"+" <Condition> <Type_1_Interaction>}
```

Figure 12-10   Rule 8

12-10 Type_2 Expression is either an Unconditional Type_2 Interaction or a Conditional Type_2 Interaction

Rule 9 describes that the type_2 expression (i.e. Type_2_Expression) is either an unconditional type_2 interaction (i.e. Type_2_Interaction) or a conditional type_2 interaction (i.e. one-or-more-armed conditional expression of Type_2_Interaction), as shown in Figure 12-11.

Rule 9

<Type_2_Expression>   ::=   <Type_2_Interaction>
  | <Condition> <Type_2_Interaction>
      {"+" <Condition> <Type_2_Interaction>}

Figure 12-11   Rule 9

12-11 An Actor Interacting with a Component Defines the Type_1 Interaction

Rule 10 describes that an actor interacting with a component defines the type_1 interaction, as shown in Figure 12-12.

Rule 10
<Type_1_Interaction> ::= <Actor> <Operation_Call_Or_Return>   <Operation_Call_Or_Return_Formula> <Component>

Figure 12-12   Rule 10

12-12 A Component Interacting with another Component Defines the Type_2 Interaction

Rule 11 describes that a component interacting with another component defines the type_2 interaction, as shown in Figure 12-13.

Rule 11
<Type_2_Interaction> ::= <Component> <Operation_Call_Or_Return> <Operation_Call_Or_Return_Formula> <Component>

Figure 12-13　Rule 11

Chapter 13: Multi-Queue SBC Process Algebra Transitional Semantics Regarding the Multi-Level View

In the chapter, we illustrate in detail those multi-queue SBC process algebra transitional semantics which regards the multi-level view.

13-1 Transitional Semantics for the Multi-Level View

As shown in Figure 13-1, we assume an infinite set Δ of type_1_or_2 interactions, and use a_1, a_2...to range over Δ. Henceforward we let $\Omega = \Delta \cup \{\lambda\}$, the set of all possible interactions, and use α_1, α_2...to to range over Ω. Further, we let X be the set of process variables, and use X_1, X_2...to range over X. We let Φ be the set of process Constants, and use A_1, A_2...to range over Φ. We let Π be the set of processes, and use P_1, Q_1...to range over Π. We let Ψ be the set of process expressions, and use E_1, E_2...to range over Ψ. We let Γ be the set of components, and use C_1, C_2...to range over Γ.

Entity set	Entity name	Type of entity
Δ	a_1, a_2...	type_1_or_2 interactions
	λ	internal interaction
Ω	α_1, α_2,...	type_1_or_2 or internal interactions
$\Omega *$	s_1, s_2...	interaction sequences
	f_1, f_2...	renaming functions
X	X_1, X_2...	process variables
Φ	A_1, A_2...	process Constants
Π	P_1, Q_1...	processes
Ψ	E_1, E_2...	process expressions
Γ	C_1, C_2...	components
	S	bisimulations

Figure 13-1   Entities

In giving meaning to the multi-queue SBC process algebra for the multi-level view, we shall use the following labelled transition system (LTS) [Miln89, Miln99]

$$(\Psi, \Omega, \rightarrow)$$

which consists of a set Ψ of process expressions, a set Ω of "type_1_or_2 or internal interactions", and a transition relation $\rightarrow \subseteq \Psi \times \Omega \times \Psi$ where $(E_i, \alpha, E_j) \in \rightarrow$ is denoted by $E_i \xrightarrow{\alpha} E_j$.

The semantics for Ψ consists in the transition rules of each transition relation $\rightarrow$ over $\Psi \times \Omega \times \Psi$. These transition rules will follow the structure of expressions.

As shown in Figure 13-2, we give the complete set of transition rules; the names Prefix, Parallel, Recursion, Rename, and Constant indicate that the rules are associated respectively with Prefix, Parallel Composition, Recursion, and Structural Composition and with Constants.

Prefix	$$\dfrac{}{\alpha \bullet E \xrightarrow{\alpha} E}$$
Parallel$_1$	$$\dfrac{E \xrightarrow{\alpha} E'}{E \parallel F \xrightarrow{\alpha} E' \parallel F}$$
Parallel$_2$	$$\dfrac{F \xrightarrow{\alpha} F'}{E \parallel F \xrightarrow{\alpha} E \parallel F'}$$
Recursion	$$\dfrac{\mathbf{fix}(X=z\{\mathbf{fix}(X=z)/X\}) \xrightarrow{\alpha} E'}{\mathbf{fix}(X=z) \xrightarrow{\alpha} E'}$$
Rename	$$\dfrac{E \xrightarrow{\alpha} E'}{E[f] \xrightarrow{\alpha[f]} E'[f]}$$
Constant	$$\dfrac{P \xrightarrow{\alpha} P'}{A \xrightarrow{\alpha} P'} \quad (A \overset{\text{def}}{=} P)$$

Figure 13-2   Transition Rules for the Multi-Level View

13-2 Rule of Prefix

The rule for Prefix, shown in Figure 13-3, can be read as follows: Under any circumstances, we always infer $\alpha \bullet E \xrightarrow{\alpha} E$. That is, an expression, with an interaction prefixed to it, will use this interaction to accomplish the transition.

$$\frac{\rule{3cm}{0.4pt}}{\alpha \bullet E \xrightarrow{\alpha} E}$$

Figure 13-3   Rule of Prefix

13-3 Rules of Parallel Composition

There are two transition rules for parallel composition. Rule Parallel$_1$, as shown in Figure 13-4, indicates that from $E \xrightarrow{\alpha} E'$ we shall infer $E \parallel F \xrightarrow{\alpha} E' \parallel F$.

$$\frac{E \xrightarrow{\alpha} E'}{E \parallel F \xrightarrow{\alpha} E' \parallel F}$$

Figure 13-4   Rule Parallel$_1$

Rule Parallel$_2$, as shown in Figure 13-5, indicates that from $F \xrightarrow{\alpha} F'$ we shall infer $E \parallel F \xrightarrow{\alpha} E \parallel F'$.

$$\frac{F \xrightarrow{\alpha} F'}{E \parallel F \xrightarrow{\alpha} E \parallel F'}$$

Figure 13-5   Rule Parallel$_2$

13-4 Rule of Recursion

The rule for Recursion, shown in Figure 13-6, can be read as follows: This says that any interaction which may be inferred for the **fix** expression 'unwound' once (by substituting itself for its bound variable) may be inferred for the **fix** expression itself.

$$\frac{\mathbf{fix}(X=z\{\mathbf{fix}(X=z)/X\}) \xrightarrow{\alpha} E\,'}{\mathbf{fix}(X=z) \xrightarrow{\alpha} E\,'}$$

Figure 13-6   Rule of Recursion

13-5 Rule of Rename

The rule for Rename, shown in Figure 13-7, can be read as follows: Rule Rename indicates that from $E \xrightarrow{\alpha} E\,'$ we shall infer $E[f] \xrightarrow{\alpha[f]} E\,'[f]$.

$$\frac{E \xrightarrow{\alpha} E\,'}{E[f] \xrightarrow{\alpha[f]} E\,'[f]}$$

Figure 13-7   Rule of Structural Composition

136

13-6 Rule of Constants

The rule for Constants, shown in Figure 13-8, can be read as follows: the rule of Constants asserts that each Constant has the same transitions as its defining expression.

$$\frac{P \xrightarrow{\alpha} P'}{A \xrightarrow{\alpha} P'} \quad (A \stackrel{\text{def}}{=} P)$$

Figure 13-8   Rule of Constants

Chapter 14: Theory of Observation Congruence

The two processes are said to be observation congruent if their observational behaviors are the same, but internal behaviors may differ widely [Miln89]. In this chapter, we introduce the theory of observation congruence. First, we give some preliminary definitions. We then define the observation equivalence. Last, we define the observation congruence.

14-1 Preliminary Definitions

A few preliminary definitions are needed. Definition 14.1 and definition 14.2 are easy to comprehend.

Definition 14.1 $\quad \widehat{\alpha} = \varepsilon$ (the empty sequence), if $\alpha = \lambda$

$\qquad\qquad\qquad = \alpha$, if $\alpha \neq \lambda$

Definition 14.2 $\quad E \overset{\alpha}{\Longrightarrow} E' \quad$ iff $\quad E \, (\overset{\lambda}{\rightarrow})^* \overset{\alpha}{\Longrightarrow} (\overset{\lambda}{\rightarrow})^* E'$

$\qquad\qquad\quad E \overset{\varepsilon}{\Longrightarrow} E' \quad$ iff $\quad E \, (\overset{\lambda}{\rightarrow})^* (\overset{\lambda}{\rightarrow})^* E'$

14-2 Observation Equivalence

To achieve the observation congruence, we need to define the observation equivalence first. Definition 14.3 and definition 14.4 together define the observation equivalence.

Definition 14.3   A binary relation $S \subseteq \Pi \times \Pi$ over multi-queue processes is a *bisimulation* if $(P, Q) \in S$ implies, for all $\alpha \in \Omega$,

(i) Whenever $P \xrightarrow{\alpha} P'$ then, for some Q', $Q \overset{\widehat{\alpha}}{\Longrightarrow} Q'$ and $(P', Q') \in S$

(ii) Whenever $Q \xrightarrow{\alpha} Q'$ then, for some P', $P \overset{\widehat{\alpha}}{\Longrightarrow} P'$ and $(P', Q') \in S$

Definition 14.4 P and Q are observation equivalent, written $P \approx Q$, if $(P, Q) \in S$ for some bisimulation S. That is,

$$\approx = \bigcup \; (S : S \text{ is a bisimulation})$$

14-3 Observation Congruence

Once we have the definition of observation equivalence, i.e. $\overset{\approx}{}$, we shall use it to define the observation congruence.

Definition 14.5 P and Q are observation congruent,
written $P = Q$, if for all α

(i) Whenever $P \overset{\alpha}{\longrightarrow} P'$ then, for some Q', $Q \overset{\alpha}{\Longrightarrow} Q'$ and $P' \approx Q'$;

(ii) Whenever $Q \overset{\alpha}{\longrightarrow} Q'$ then, for some P', $P \overset{\alpha}{\Longrightarrow} P'$ and $P' \approx Q'$.

140

Chapter 15: Kurdi University as an Example

In the SBC view model of *Kurdi University* as shown in Figure 15-1, dimension 1 stands for the evolution&motivation view which contains the strategy/version 1, strategy/version 2, strategy/version 3, strategy/version 4,…, and strategy/version ∞ views; dimension 2 stands for the multi-level view which contains the concept, analysis, design, and implementation views; dimension 3 stands for the systemic view which contains the collection of all interaction flow diagrams.

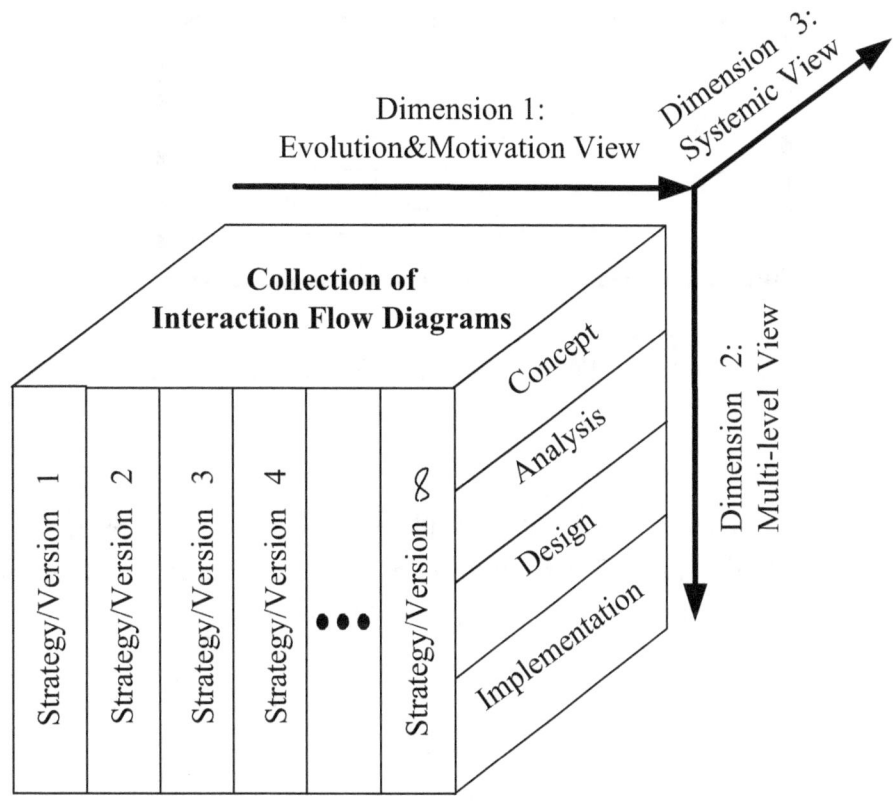

Figure 15-1   SBC View Model

In the SBC multi-level view of *Kurdi University*, concept view is one level up structural composition (with observation congruence verification) of the analysis view; analysis view is one level up structural composition (with observation congruence verification) of the design view; design view is one level up structural composition (with observation congruence verification) of the implementation view.

15-1 Multi-Queue SBC Process of the Kurdi University's Concept View

We draw the Architecture Hierarchy Diagram (AHD) of the concept view of *Kurdi University* as shown in Figure 15-2.

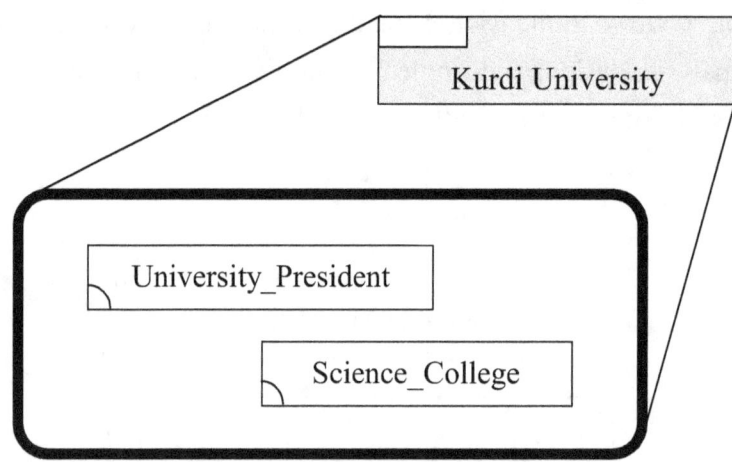

Figure 15-2   AHD of the Concept View of *Kurdi University*

The overall behavior of the concept view of *Kurdi University* includes two behaviors: *Study_Calculus_Course* and *Study_Algebra_Course* as shown in Figure 15-3. Each of them is described by an individual IFD.

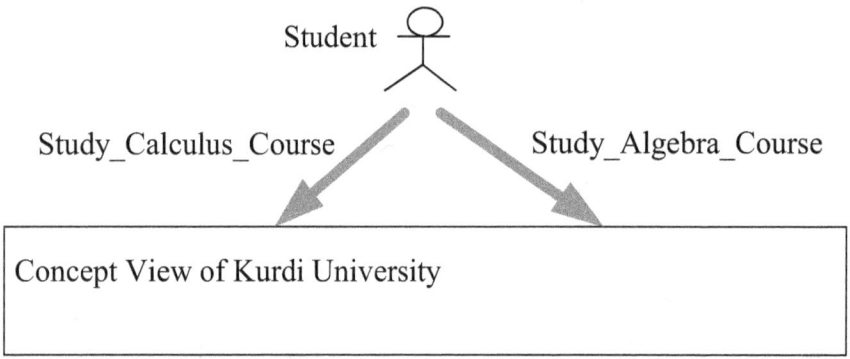

Figure 15-3   Overall Behavior of the Concept View
of Kurdi University

Figure 15-4 shows the concept view's IFD of the *Study_Calculus_Course* behavior. First, actor *Student* interacts with the *University_President* component through the *University_Teach_Calculus* operation call interaction. Next, component

University_President interacts with the *Science_College* component through the *College_Teach_Calculus* operation call interaction.

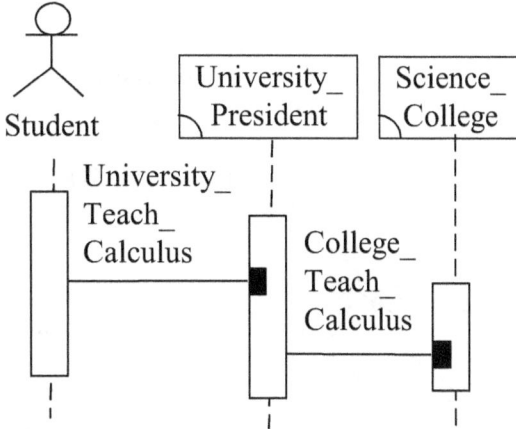

Figure 15-4   Concept View's IFD of the *Study_Calculus_Course* Behavior

Figure 15-5 shows the concept view's IFD of the *Study_Algebra_Course* behavior. First, actor *Student* interacts with the *University_President* component through the *University_Teach_Algebra* operation call interaction. Next, component *University_President* interacts with the *Science_College* component through the *College_Teach_Algebra* operation call interaction.

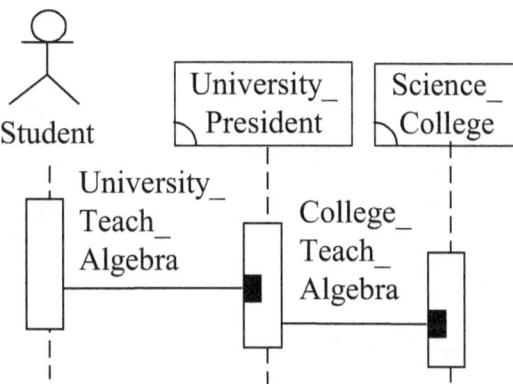

Figure 15-5   Concept View's IFD of the *Study_Algebra_Course* Behavior

144

We draw the multi-queue SBC process algebra Backus-Naur Form tree of the concept view of *Kurdi University* as shown in Figure 15-6.

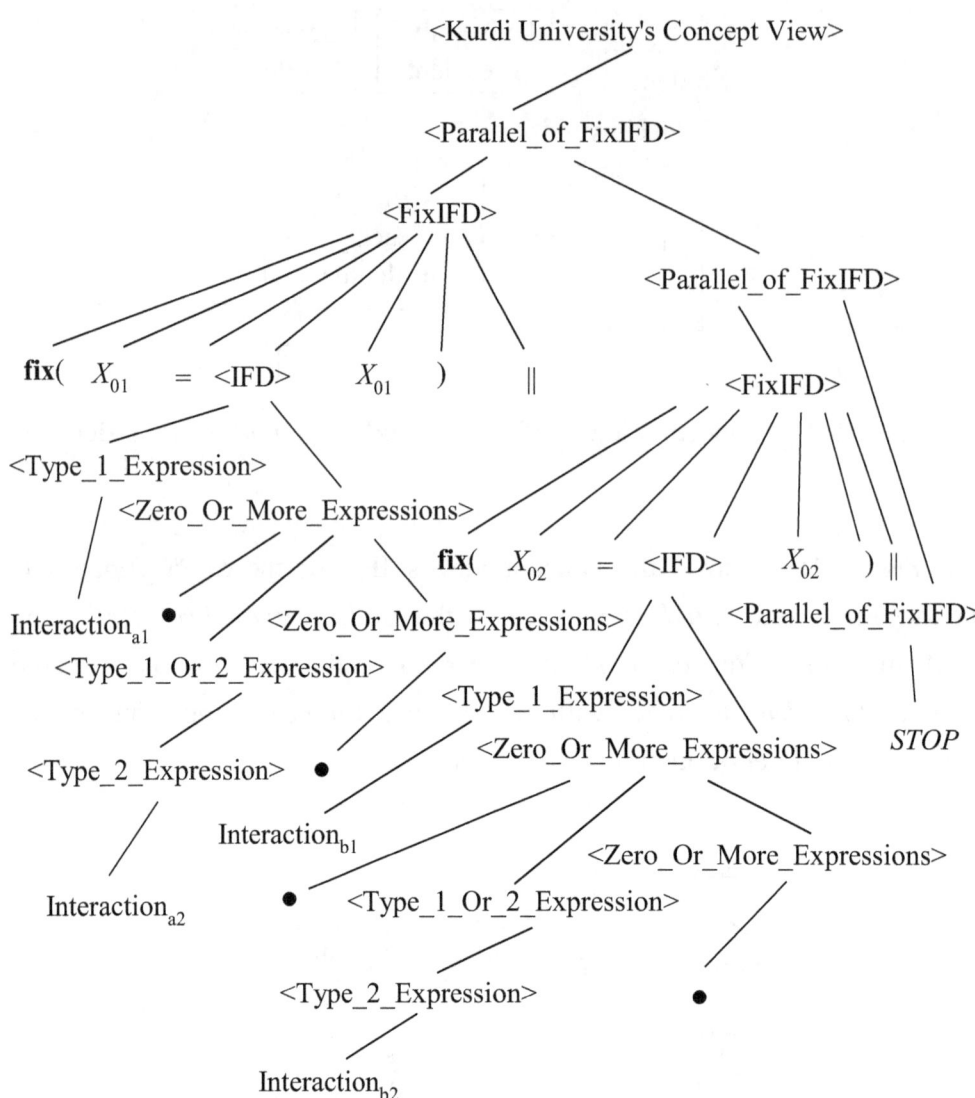

Figure 15-6 M-SBC-PA Backus-Naur Form Tree of the Concept View of Kurdi University

Interaction$_{a1}$ stands for the 1st interaction of the ath interaction flow diagram of the concept view of *Kurdi University*, as shown in Figure 15-7. Interaction$_{a1}$ is a type_1 interaction which describes the *Student* actor interacts with the *University_President* component.

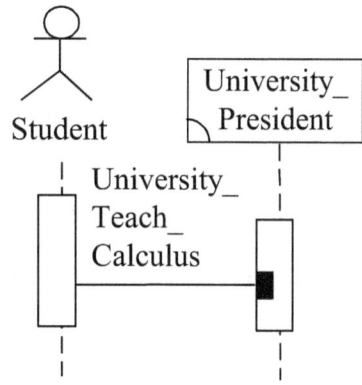

Figure 15-7   Interaction$_{a1}$

Interaction$_{a2}$ stands for the 2nd interaction of the ath interaction flow diagram of the concept view of *Kurdi University*, as shown in Figure 15-8. Interaction$_{a2}$ is a type_2 interaction which describes the *University_President* component interacts with the *Science_College* component.

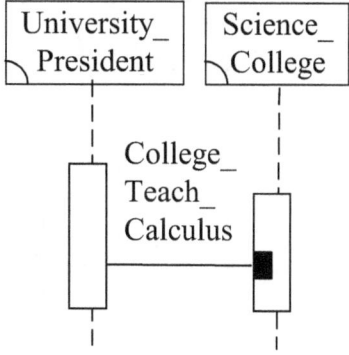

Figure 15-8   Interaction$_{a2}$

Interaction$_{b1}$ stands for the 1st interaction of the bth interaction flow diagram of the concept view of *Kurdi University*, as shown in Figure 15-9. Interaction$_{b1}$ is a type_1 interaction which describes the *Student* actor interacts with the *University_President* component.

146

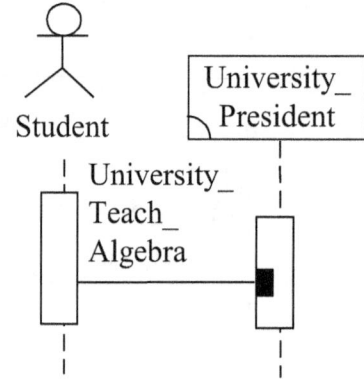

Figure 15-9   Interaction$_{b1}$

Interaction$_{b2}$ stands for the 2nd interaction of the bth interaction flow diagram of the concept view of *Kurdi University*, as shown in Figure 15-10. Interaction$_{b2}$ is a type_2 interaction which describes the *University_President* component interacts with the *Science_College* component.

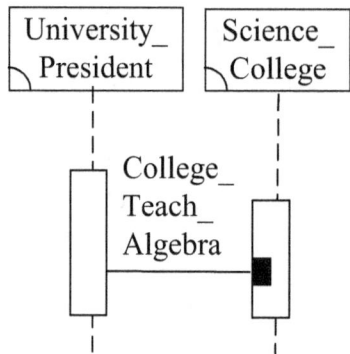

Figure 15-10   Interaction$_{b2}$

FixIFD$_a$ describes the recursion of the ath interaction flow diagram, i.e. *Study_Calculus_Course* behavior, of the concept view of *Kurdi University*. FixIFD$_a$ is syntactically represented as "**fix**(X_{01}=Interaction$_{a1}$●Interaction$_{a2}$●X_{01})" as shown in Figure 15-11.

FixIFD$_a$ $\overset{def}{=\!=}$

$\mathbf{fix}(X_{01}=\text{Interaction}_{a1}\bullet \text{Interaction}_{a2}\bullet X_{01})$

Figure 15-11   FixIFD$_a$

FixIFD$_b$ describes the recursion of the bth interaction flow diagram, i.e. *Study_Algebra_Course* behavior, of the concept view of *Kurdi University*. FixIFD$_b$ is syntactically represented as "$\mathbf{fix}(X_{02}=\text{Interaction}_{b1}\bullet\text{Interaction}_{b2}\bullet X_{02})$" as shown in Figure 15-12.

FixIFD$_b$ $\overset{def}{=\!=}$

$\mathbf{fix}(X_{02}=\text{Interaction}_{b1}\bullet \text{Interaction}_{b2}\bullet X_{02})$

Figure 15-12   FixIFD$_b$

Multi-queue SBC process of the concept view of *Kurdi University* is syntactically represented as FixIFD$_a$‖FixIFD$_b$ which equals to "$\mathbf{fix}(X_{01}=\text{Interaction}_{a1}\bullet\text{Interaction}_{a2}\bullet X_{01})$ ‖ $\mathbf{fix}(X_{02}=\text{Interaction}_{b1}\bullet\text{Interaction}_{b2}\bullet X_{02})$" as shown in Figure 15-13.

Kurdi University's Concept View $\overset{def}{=\!=}$

$\mathbf{fix}(X_{01}=\text{Interaction}_{a1}\bullet \text{Interaction}_{a2}\bullet X_{01})$ ‖
$\mathbf{fix}(X_{02}=\text{Interaction}_{b1}\bullet \text{Interaction}_{b2}\bullet X_{02})$

Figure 15-13   Kurdi University's Concept View

148

15-2 Multi-Queue SBC Process of the Kurdi University's Analysis View

We draw the Architecture Hierarchy Diagram (AHD) of the analysis view of *Kurdi University* as shown in Figure 15-14.

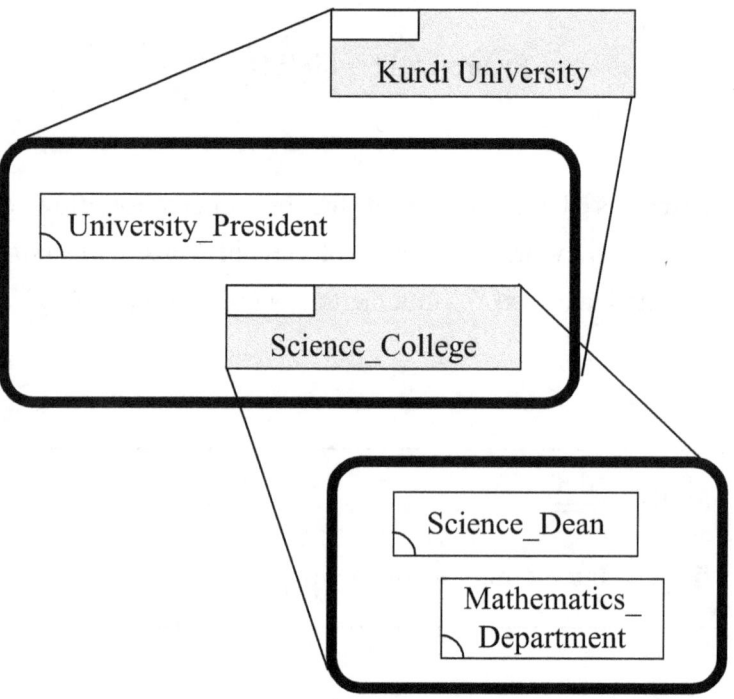

Figure 15-14   AHD of the Analysis View of *Kurdi University*

The overall behavior of the analysis view of *Kurdi University* includes two behaviors: *Study_Calculus_Course* and *Study_Algebra_Course* as shown in Figure 15-15. Each of them is described by an individual IFD.

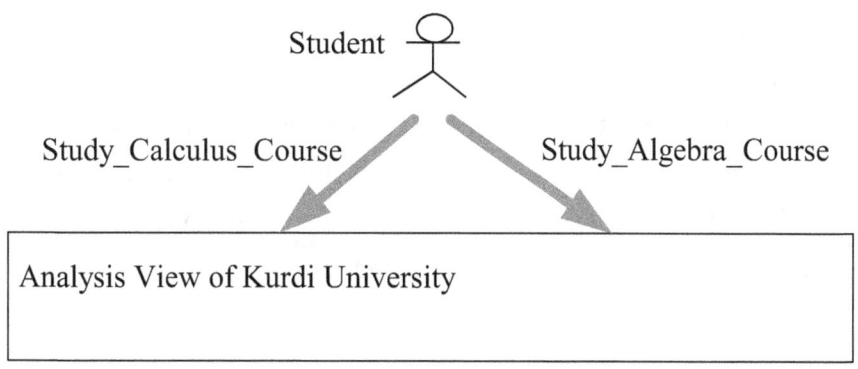

Figure 15-15   Overall Behavior of the Analysis View
of Kurdi University

Figure 15-16 shows the analysis view's IFD of the *Study_Calculus_Course* behavior. First, actor *Student* interacts with the *University_President* component through the *University_Teach_Calculus* operation call interaction. Next, component *University_President* interacts with the *Science_Dean* component through the *College_Teach_Calculus* operation call interaction. Finally, component *Science_Dean* interacts with the *Mathematics_Department* component through the *Department_Teach_Calculus* operation call interaction.

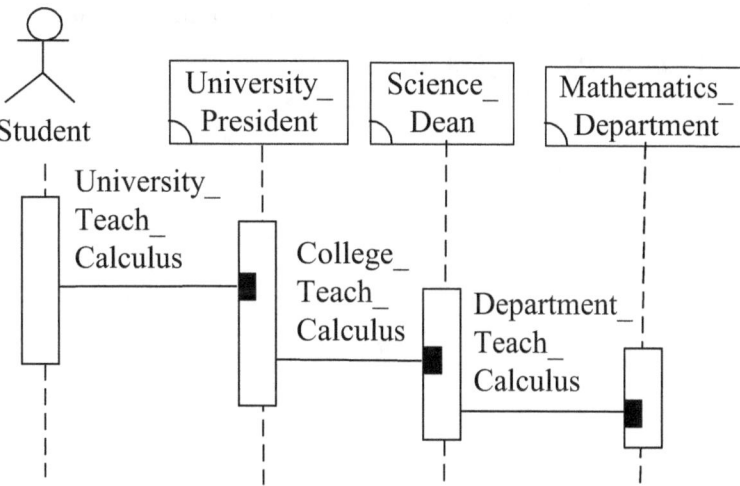

Figure 15-16   Analysis View's IFD of the *Study_Calculus_Course* Behavior

Figure 15-17 shows the analysis view's IFD of the *Study_Algebra_Course* behavior. First, actor *Student* interacts with the *University_President* component through the *University_Teach_Algebra* operation call interaction. Next, component *University_President* interacts with the *Science_Dean* component through the *College_Teach_Algebra* operation call interaction. Finally, component *Science_Dean* interacts with the *Mathematics_Department* component through the *Department_Teach_Algebra* operation call interaction.

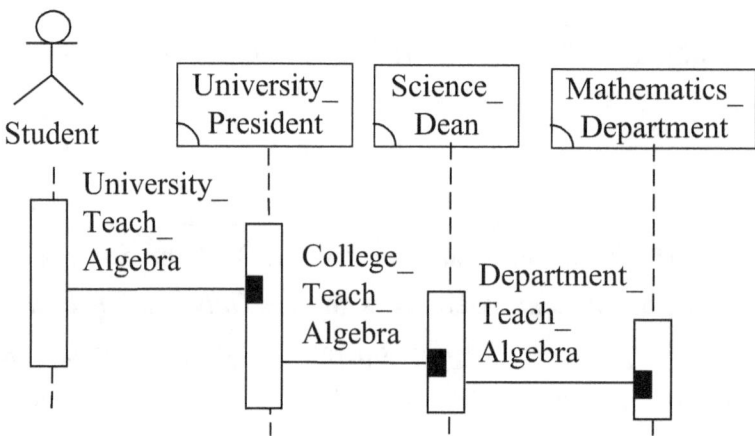

Figure 15-17   Analysis View's IFD of the *Study_Algebra_Course* Behavior

We draw the multi-queue SBC process algebra Backus-Naur Form tree of the analysis view of *Kurdi University* as shown in Figure 15-18.

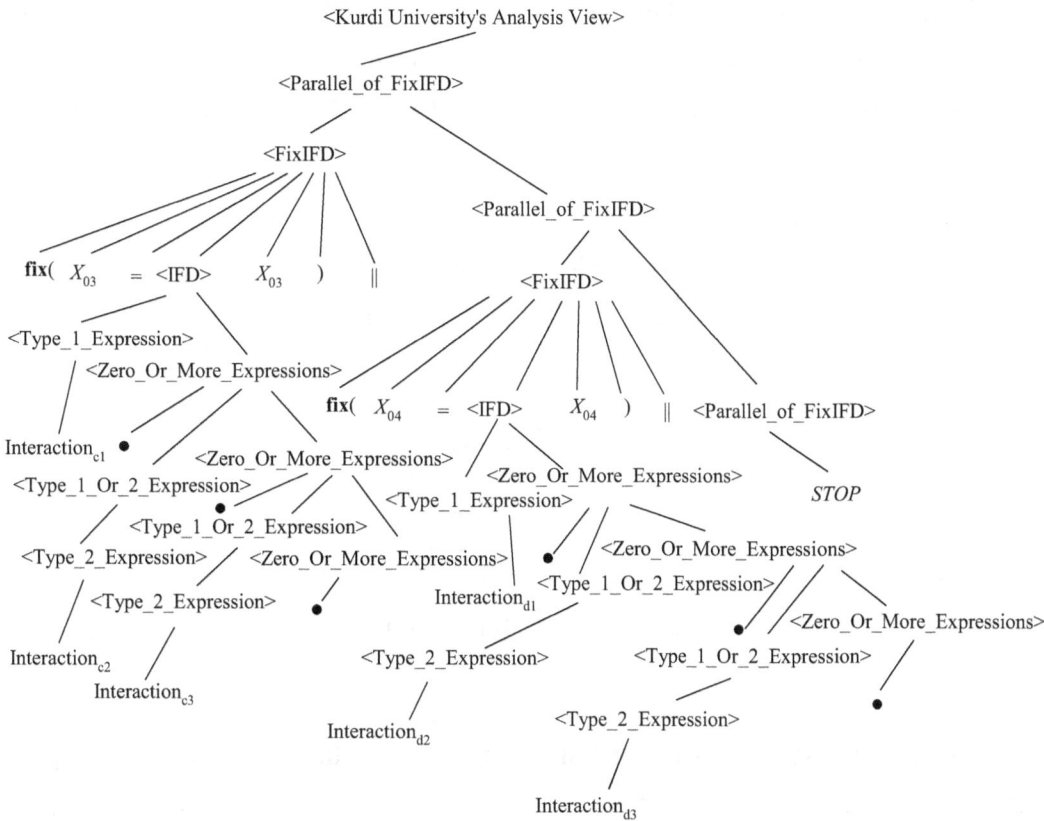

Figure 15-18   M-SBC-PA Backus-Naur Form Tree of the Analysis View of Kurdi University

Interaction$_{c1}$ stands for the 1st interaction of the cth interaction flow diagram of the analysis view of *Kurdi University*, as shown in Figure 15-19. Interaction$_{c1}$ is a type_1 interaction which describes the *Student* actor interacts with the *University_President* component.

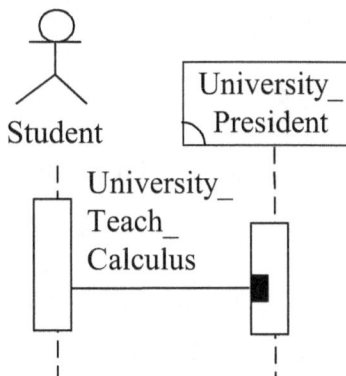

Figure 15-19   Interaction$_{c1}$

Interaction$_{c2}$ stands for the 2nd interaction of the cth interaction flow diagram of the analysis view of *Kurdi University*, as shown in Figure 15-20. Interaction$_{c2}$ is a type_2 interaction which describes the *University_President* component interacts with the *Science_Dean* component.

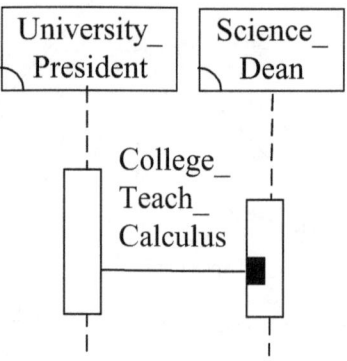

Figure 15-20　　Interaction$_{c2}$

Interaction$_{c3}$ stands for the 3rd interaction of the cth interaction flow diagram of the analysis view of *Kurdi University*, as shown in Figure 15-21. Interaction$_{c3}$ is a type_2 interaction which describes the *Science_Dean* component interacts with the *Mathematics_Department* component.

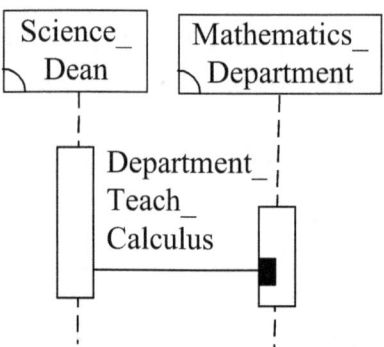

Figure 15-21　　Interaction$_{c3}$

Interaction$_{d1}$ stands for the 1st interaction of the dth interaction flow diagram of the analysis view of *Kurdi University*, as shown in Figure 15-22. Interaction$_{d1}$ is a type_1 interaction which describes the *Student* actor interacts with the *University_President* component.

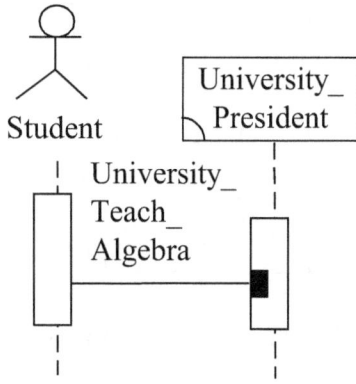

Figure 15-22   Interaction$_{d1}$

Interaction$_{d2}$ stands for the 2nd interaction of the dth interaction flow diagram of the analysis view of *Kurdi University*, as shown in Figure 15-23.   Interaction$_{d2}$ is a type_2 interaction which describes the *University_President* component interacts with the *Science_Dean* component.

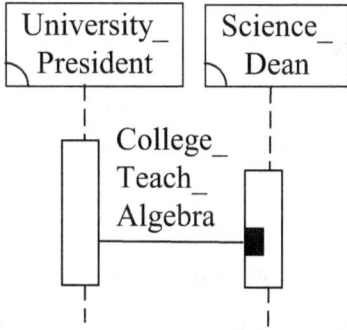

Figure 15-23   Interaction$_{d2}$

Interaction$_{d3}$ stands for the 3rd interaction of the dth interaction flow diagram of the analysis view of *Kurdi University*, as shown in Figure 15-24. Interaction$_{d3}$ is a type_2 interaction which describes the *Science_Dean* component interacts with the *Mathematics_Department* component.

154

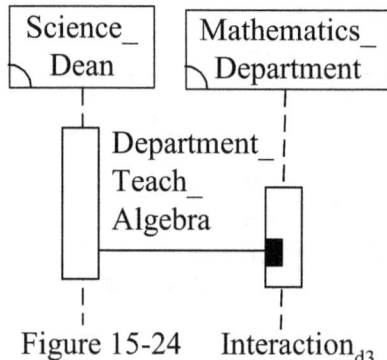

Figure 15-24   Interaction$_{d3}$

FixIFD$_c$ describes the recursion of the cth interaction flow diagram, i.e. *Study_Calculus_Course* behavior, of the analysis view of *Kurdi University*. FixIFD$_c$ is syntactically represented as "**fix**(X_{03}=Interaction$_{c1}$●Interaction$_{c2}$●Interaction$_{c3}$●X_{03})", as shown in Figure 15-25.

$$\text{FixIFD}_c \overset{\text{def}}{=\!=\!=}$$
$$\textbf{fix}(X_{03} = \text{Interaction}_{c1} ● \ \text{Interaction}_{c2} ● \ \text{Interaction}_{c3} ● \ X_{03})$$

Figure 15-25   FixIFD$_c$

FixIFD$_d$ describes the recursion of the dth interaction flow diagram, i.e. *Study_Algebra_Course* behavior, of the analysis view of *Kurdi University*. FixIFD$_d$ is syntactically represented as "**fix**(X_{04}=Interaction$_{d1}$●Interaction$_{d2}$●Interaction$_{d3}$●X_{04})", as shown in Figure 15-26.

$$\text{FixIFD}_d \overset{\text{def}}{=\!=\!=}$$
$$\textbf{fix}(X_{04} = \text{Interaction}_{d1} ● \ \text{Interaction}_{d2} ● \ \text{Interaction}_{d3} ● \ X_{04})$$

Figure 15-26   FixIFD$_d$

Multi-queue SBC process of the analysis view of *Kurdi University* is syntactically represented as $FixIFD_c \| FixIFD_d$ which equals to "$\mathbf{fix}(X_{03}=Interaction_{c1} \bullet Interaction_{c2} \bullet Interaction_{c3} \bullet X_{03})$ $\|$ $\mathbf{fix}(X_{04}=Interaction_{d1} \bullet Interaction_{d2} \bullet Interaction_{d3} \bullet X_{04})$" as shown in Figure 15-27.

Kurdi University's Analysis View $\overset{def}{=\!=}$

$\mathbf{fix}(X_{03}=Interaction_{c1} \bullet Interaction_{c2} \bullet Interaction_{c3} \bullet X_{03})$ $\|$
$\mathbf{fix}(X_{04}=Interaction_{d1} \bullet Interaction_{d2} \bullet Interaction_{d3} \bullet X_{04})$

Figure 15-27   Kurdi University's Analysis View

15-3 Multi-Queue SBC Process of the Structural Composition of the Kurdi University's Analysis View

Structural composition of the analysis view of *Kurdi University* means to compose the *Science_Dean* and *Mathematics_Department* components into the *Science_College* component. That is, we will rename the *Science_Dean* component to the *Science_College* component; we also will rename the *Mathematics_Department* component to the *Science_College* component, as shown in Figure 15-28.

[Science_College/Science_Dean,
Science_College/Mathematics_Department]

Figure 15-28   Rename the *Science_Dean*,
Mathematics_Department Components to the *Science_College*
Component

We draw the multi-queue SBC process algebra Backus-Naur Form tree of the structural composition of the analysis view of *Kurdi University* as shown in Figure 15-29.

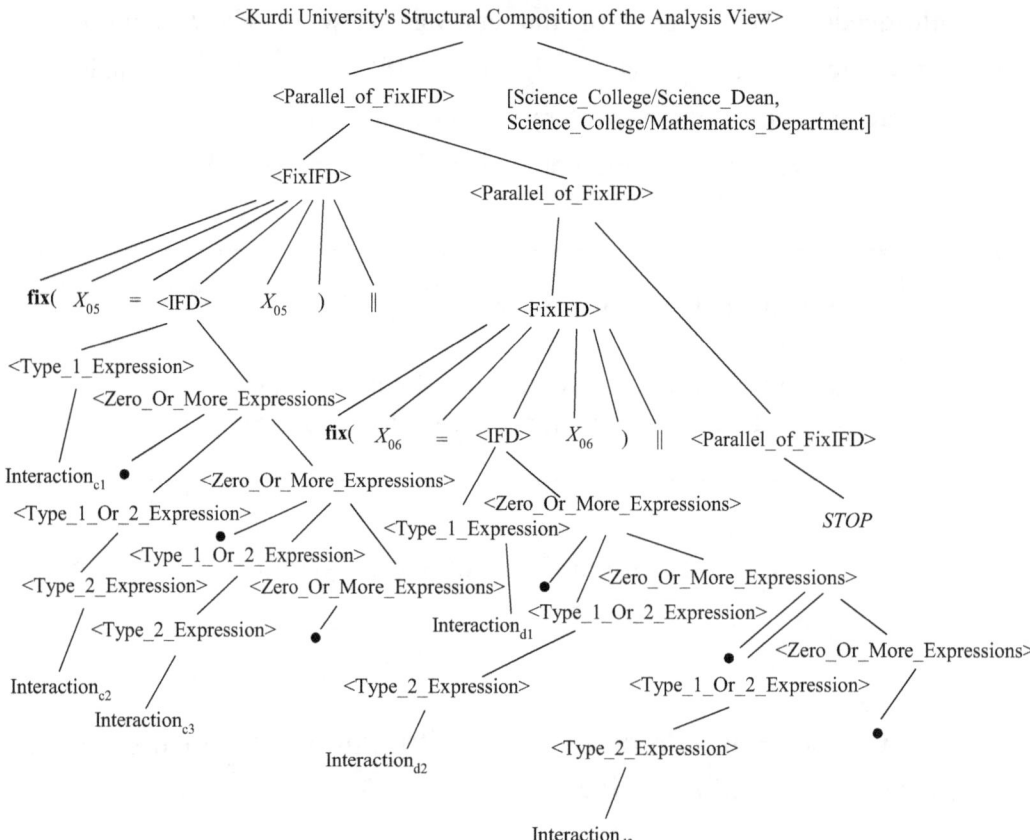

Figure 15-29   M-SBC-PA Backus-Naur Form Tree of the Structural Composition of the Analysis View of Kurdi University

Interaction$_{e1}$=Interaction$_{c1}$[Science_College/Science_Dean,Science_College/ Mathematics_Department]=Interaction$_{a1}$ stands for the 1st interaction of the eth interaction flow diagram of the structural composition of the analysis view of *Kurdi University*, as shown in Figure 15-30. Interaction$_{e1}$ is a type_1 interaction which describes the *Student* actor interacts with the *University_President* component.

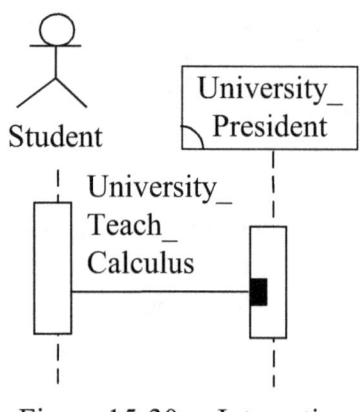

Figure 15-30   Interaction$_{e1}$

Interaction$_{e2}$=Interaction$_{c2}$[Science_College/Science_Dean,Science_College/ Mathematics_Department]=Interaction$_{a2}$ stands for the 2nd interaction of the eth interaction flow diagram of the structural composition of the analysis view of *Kurdi University*, as shown in Figure 15-31. Interaction$_{e2}$ is a type_2 interaction which describes the *University_President* component interacts with the *Science_College* component.

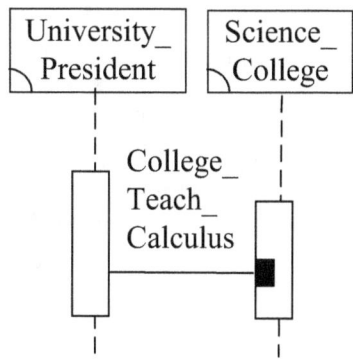

Figure 15-31   Interaction$_{e2}$

Interaction$_{e3}$=Interaction$_{c3}$[Science_College/Science_Dean,Science_College/ Mathematics_Department] stands for the 3rd interaction of the eth interaction flow diagram of the structural composition of the analysis view of *Kurdi University*, as shown in Figure 15-32. Interaction$_{e3}$ is an internal interaction (i.e. λ) inside the *Science_College* component.

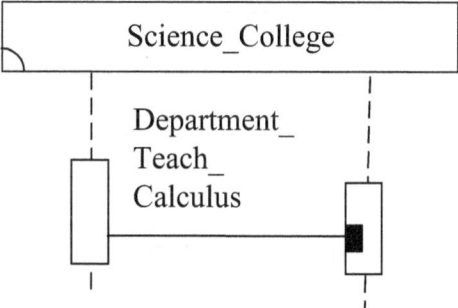

Figure 15-32   Interaction$_{e3}$

Interaction$_{f1}$=Interaction$_{d1}$[Science_College/Science_Dean,Science_College/ Mathematics_Department]=Interaction$_{b1}$ stands for the 1st interaction of the fth interaction flow diagram of the structural composition of the analysis view of *Kurdi University*, as shown in Figure 15-33. Interaction$_{f1}$ is a type_1 interaction which describes the *Student* actor interacts with the *University_President* component.

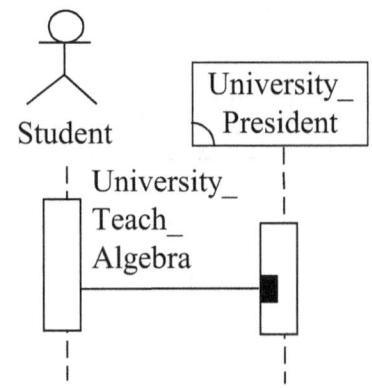

Figure 15-33   Interaction$_{f1}$

Interaction$_{f2}$=Interaction$_{d2}$[Science_College/Science_Dean,Science_College/ Mathematics_Department]=Interaction$_{b2}$ stands for the 2nd interaction of the fth interaction flow diagram of the structural composition of the analysis view of *Kurdi University*, as shown in Figure 15-34. Interaction$_{f2}$ is a type_2 interaction which describes the *University_President* component interacts with the *Science_College* component.

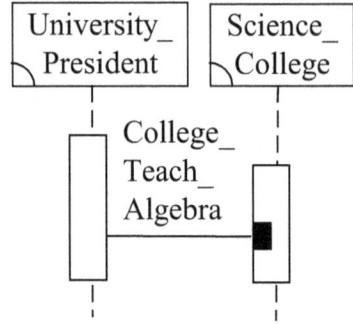

Figure 15-34   Interaction$_{f2}$

Interaction$_{f3}$=Interaction$_{d3}$[Science_College/Science_Dean,Science_College/ Mathematics_Department] stands for the 3rd interaction of the fth interaction flow diagram of the structural composition of the analysis view of *Kurdi University*, as shown in Figure 15-35. Interaction$_{f3}$ is an internal interaction (i.e. λ) inside the *Science_College* component.

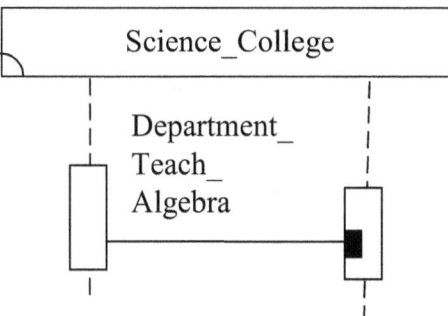

Figure 15-35   Interaction$_{f3}$

FixIFD$_e$ describes the recursion of the eth interaction flow diagram, i.e. *Study_Calculus_Course* behavior, of the structural composition of the analysis view of *Kurdi University*. FixIFD$_e$ is syntactically represented as "**fix**(X_{05}=Interaction$_{a1} \bullet$Interaction$_{a2} \bullet \lambda \bullet X_{05}$)" as shown in Figure 15-36.

$$\text{FixIFD}_e \overset{\text{def}}{=\!=}$$

$$\textbf{fix}(X_{05}=\text{Interaction}_{a1} \bullet \text{Interaction}_{a2} \bullet \lambda \bullet X_{05})$$

Figure 15-36   FixIFD$_e$

160

FixIFD$_f$ describes the recursion of the fth interaction flow diagram, i.e. *Study_Algebra_Course* behavior, of the structural composition of the analysis view of *Kurdi University*. FixIFD$_f$ is syntactically represented as "**fix**(X_{06}=Interaction$_{b1}$•Interaction$_{b2}$•λ•X_{06})" as shown in Figure 15-37.

$$\boxed{\begin{array}{l} \text{FixIFD}_f \overset{\text{def}}{=\!=} \\[2ex] \textbf{fix}(X_{06}=\text{Interaction}_{b1} \bullet \text{Interaction}_{b2} \bullet \lambda \bullet X_{06}) \end{array}}$$

Figure 15-37   FixIFD$_f$

Multi-queue SBC process of the structural composition of the analysis view of *Kurdi University* is syntactically represented as FixIFD$_e$||FixIFD$_f$ which equals to "**fix**(X_{05}=Interaction$_{a1}$•Interaction$_{a2}$•λ•X_{05}) || **fix**(X_{06}=Interaction$_{b1}$•Interaction$_{b2}$•λ•X_{06})" as shown in Figure 15-38.

$$\boxed{\begin{array}{l} \text{Kurdi University's Structural Composition of the Analysis View} \overset{\text{def}}{=\!=} \\[2ex] \textbf{fix}(X_{05}=\text{Interaction}_{a1} \bullet \text{Interaction}_{a2} \bullet \lambda \bullet X_{05}) \quad || \\ \textbf{fix}(X_{06}=\text{Interaction}_{b1} \bullet \text{Interaction}_{b2} \bullet \lambda \bullet X_{06}) \end{array}}$$

Figure 15-38   Kurdi University's Structural Composition
of the Analysis View

15-4 Observation Congruence of "the Concept View" and "the Structural Composition of the Analysis View" of Kurdi University

We syntactically represent multi-queue SBC process P_{01} as $\text{FixIFD}_a\|\text{FixIFD}_b$ which equals to "**fix**$(X_{01}=\text{Interaction}_{a1}\bullet\text{Interaction}_{a2}\bullet X_{01})$ $\|$ **fix**$(X_{02}=\text{Interaction}_{b1}\bullet\text{Interaction}_{b2}\bullet X_{02})$." The transition graph, as shown in Figure 15-39, demonstrates the semantics of process P_{01}.

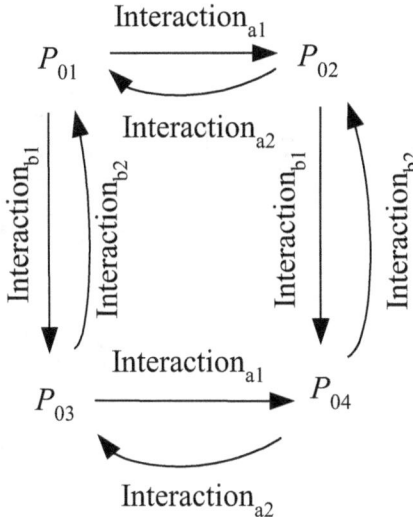

Figure 15-39   Transition Graph Demonstrates the Semantics of Process P_{01}

We syntactically represent multi-queue SBC process Q_{01} as as $\text{FixIFD}_e\|\text{FixIFD}_f$ which equals to "**fix**$(X_{05}=\text{Interaction}_{a1}\bullet\text{Interaction}_{a2}\bullet\lambda\bullet X_{05})$ $\|$ **fix**$(X_{06}=\text{Interaction}_{b1}\bullet\text{Interaction}_{b2}\bullet\lambda\bullet X_{06})$." The transition graph, as shown in Figure 15-40, demonstrates the semantics of process Q_{01}.

162

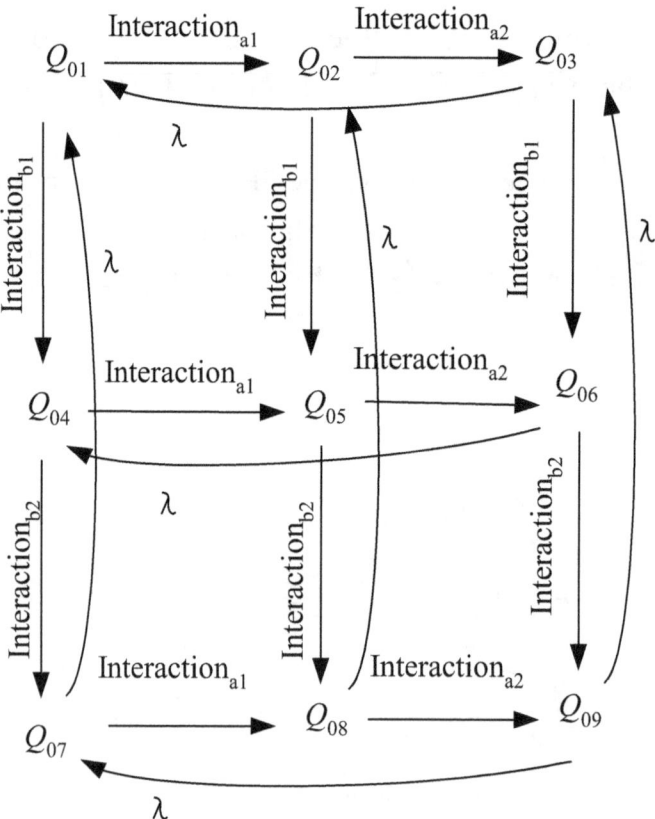

Figure 15-40   Transition Graph Demonstrates the Semantics of Process Q_{01}

We can easily verify that $S = \{(P_{01}, Q_{01}), (P_{01}, Q_{03}), (P_{01}, Q_{07}), (P_{01}, Q_{09}), (P_{02},$ $Q_{02}), (P_{02}, Q_{08}), (P_{03}, Q_{04}), (P_{03}, Q_{06}), (P_{04}, Q_{05})\}$ is a bisimulation.

Using the S bisimulation, we then are able to verify that P_{01} and Q_{01} are observation congruent because (1) $P_{01} \xrightarrow{\text{Interaction}_{a1}} P_{02}$, then we have Q_{02} that Q_{01}

$\xRightarrow{\text{Interaction}_{a1}} Q_{02}$ and $P_{02} \overset{\approx}{\sim} Q_{02}$, and (2) $P_{01} \xrightarrow{\text{Interaction}_{b1}} P_{03}$, then we have Q_{04} that

$Q_{01} \xRightarrow{\text{Interaction}_{b1}} Q_{04}$ and $P_{03} \overset{\approx}{\sim} Q_{04}$, and (3) $Q_{01} \xrightarrow{\text{Interaction}_{a1}} Q_{02}$, then we have P_{02}

that $P_{01} \xRightarrow{\text{Interaction}_{a1}} P_{02}$ and $P_{02} \overset{\approx}{\sim} Q_{02}$, and (4) $Q_{01} \xrightarrow{\text{Interaction}_{b1}} Q_{04}$, then we have

P_{03} that $P_{01} \xRightarrow{\text{Interaction}_{b1}} P_{03}$ and $P_{03} \overset{\approx}{\sim} Q_{04}$.

$P_{01} = Q_{01}$ means that "the concept view of *Kurdi University*" and "the structural composition of the analysis view of *Kurdi University*" are observation congruent.

That is, we have demonstrated that the concept view of *Kurdi University* is one level up structural composition (with observation congruence verification) of the analysis view of *Kurdi University*.

15-5 Multi-Queue SBC Process of the Kurdi University's Design View

We draw the Architecture Hierarchy Diagram (AHD) of the design view of *Kurdi University* as shown in Figure 15-41.

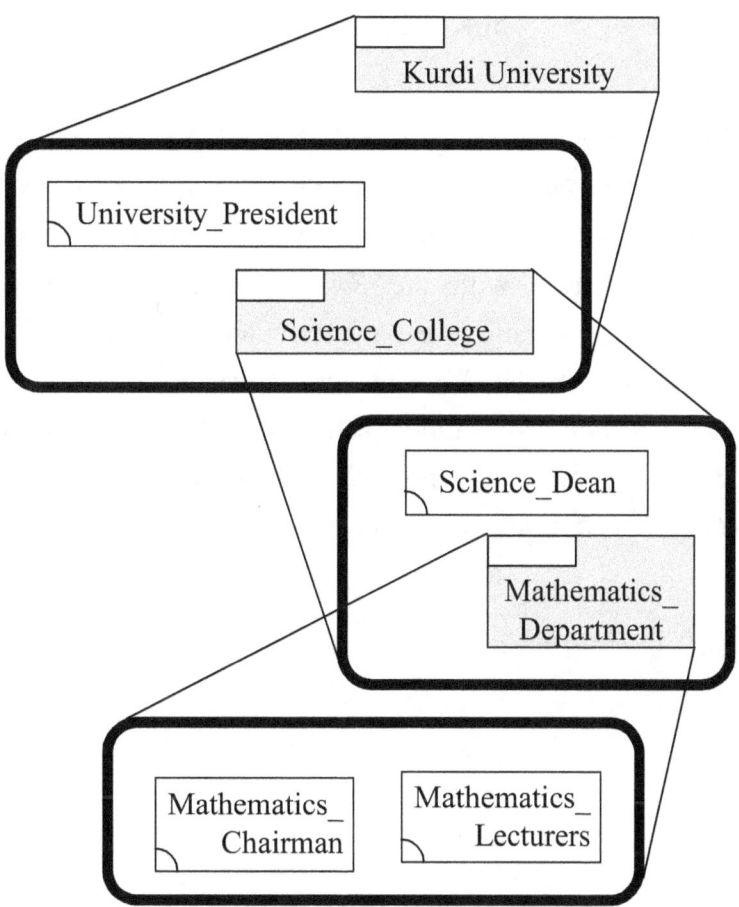

Figure 15-41   AHD of the Design View of *Kurdi University*

The overall behavior of the design view of *Kurdi University* includes two behaviors: *Study_Calculus_Course* and *Study_Algebra_Course* as shown in Figure 15-42. Each of them is described by an individual IFD.

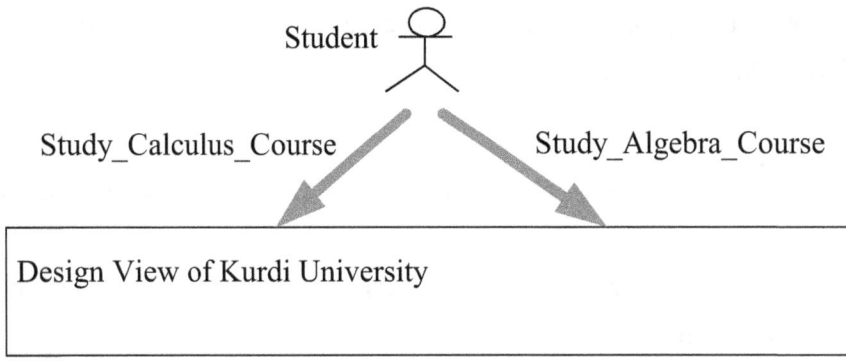

Figure 15-42   Overall Behavior of the Design View
of Kurdi University

Figure 15-43 shows the design view's IFD of the *Study_Calculus_Course* behavior. First, actor *Student* interacts with the *University_President* component through the *University_Teach_Calculus* operation call interaction. Next, component *University_President* interacts with the *Science_Dean* component through the *College_Teach_Calculus* operation call interaction. Continuingly, component *Science_Dean* interacts with the *Mathematics_Chairman* component through the *Department_Teach_Calculus* operation call interaction. Finally, component *Mathematics_Chairman* interacts with the *Mathematics_Lecturers* component through the *Lecturer_Teach_Calculus* operation call interaction.

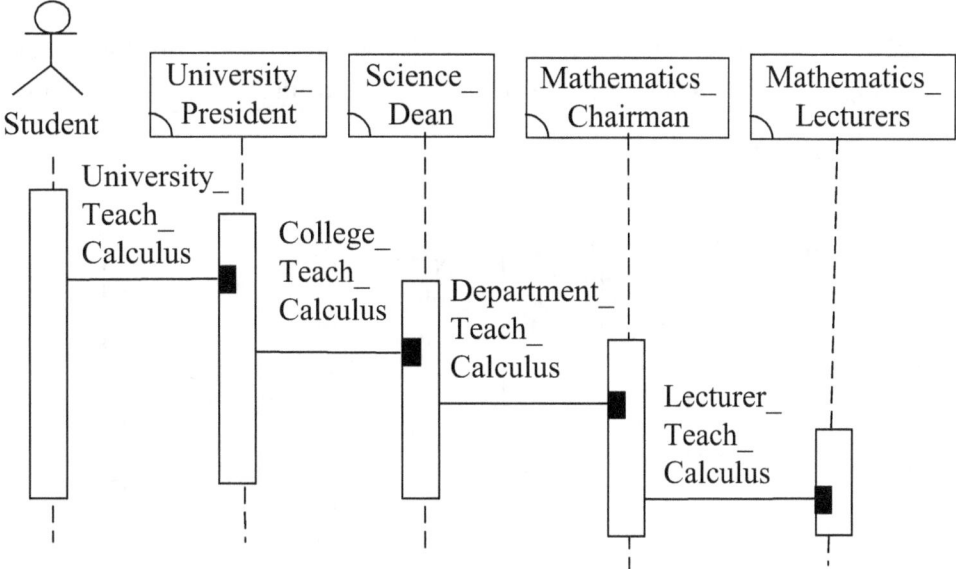

Figure 15-43   Design View's IFD of the *Study_Calculus_Course* Behavior

Figure 15-44 shows the design view's IFD of the *Study_Algebra_Course* behavior. First, actor *Student* interacts with the *University_President* component through the *University_Teach_Algebra* operation call interaction. Next, component *University_President* interacts with the *Science_Dean* component through the *College_Teach_Algebra* operation call interaction. Continuingly, component *Science_Dean* interacts with the *Mathematics_Chairman* component through the *Department_Teach_Algebra* operation call interaction. Finally, component *Mathematics_Chairman* interacts with the *Mathematics_Lecturers* component through the *Lecturer_Teach_Algebra* operation call interaction.

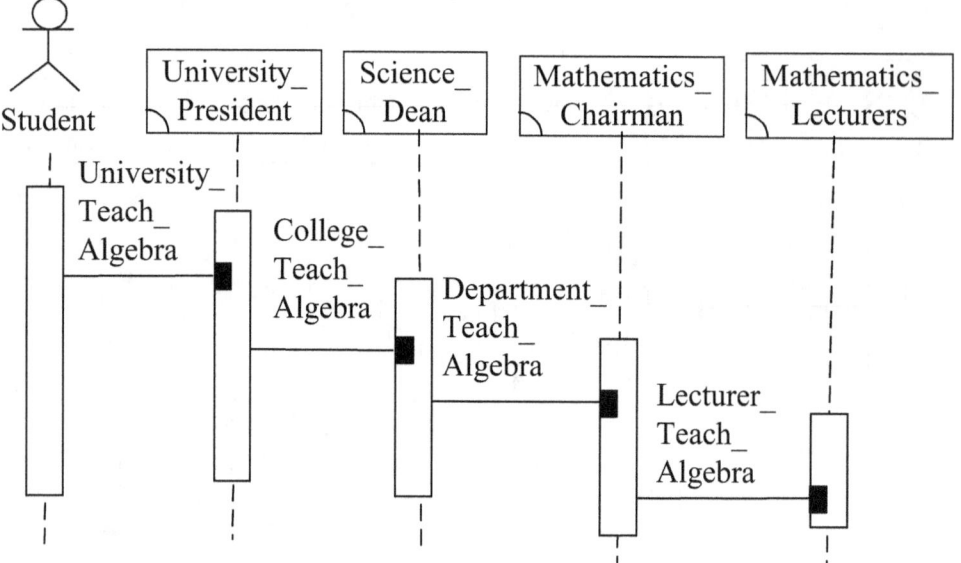

Figure 15-44   Design View's IFD of the *Study_Algebra_Course* Behavior

We draw the multi-queue SBC process algebra Backus-Naur Form tree of the design view of *Kurdi University* as shown in Figure 15-45.

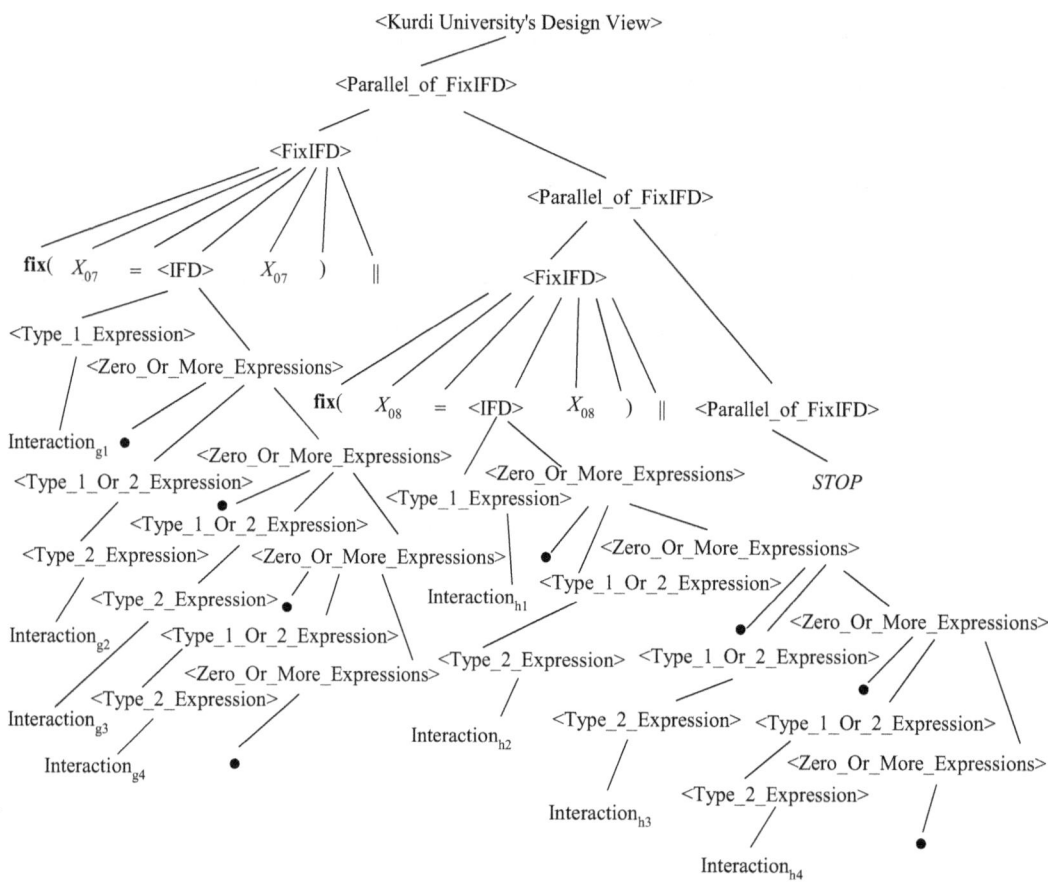

Figure 15-45   M-SBC-PA Backus-Naur Form Tree of the Design View of Kurdi University

Interaction$_{g1}$ stands for the 1st interaction of the gth interaction flow diagram of the design view of *Kurdi University*, as shown in Figure 15-46. Interaction$_{g1}$ is a type_1 interaction which describes the *Student* actor interacts with the *University_President* component.

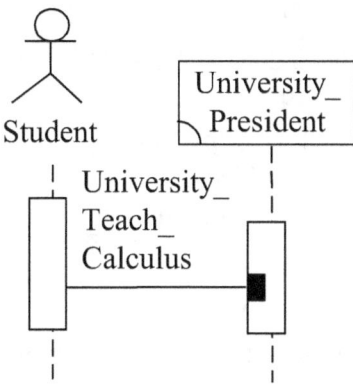

Figure 15-46   Interaction$_{g1}$

Interaction$_{g2}$ stands for the 2nd interaction of the gth interaction flow diagram of the design view of *Kurdi University*, as shown in Figure 15-47. Interaction$_{g2}$ is a type_2 interaction which describes the *University_President* component interacts with the *Science_Dean* component.

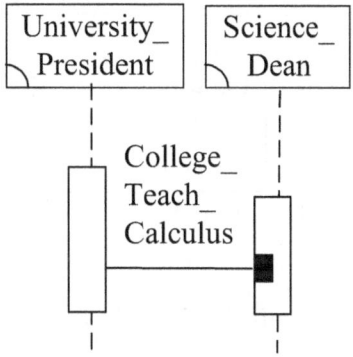

Figure 15-47   Interaction$_{g2}$

Interaction$_{g3}$ stands for the 3rd interaction of the gth interaction flow diagram of the design view of *Kurdi University*, as shown in Figure 15-48. Interaction$_{g3}$ is a type_2 interaction which describes the *Science_Dean* component interacts with the *Mathematics_Chairman* component.

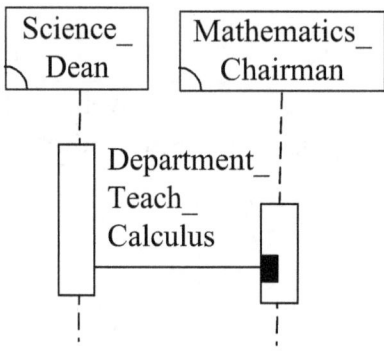

Figure 15-48   Interaction$_{g3}$

Interaction$_{g4}$ stands for the 4th interaction of the gth interaction flow diagram of the design view of *Kurdi University*, as shown in Figure 15-49. Interaction$_{g4}$ is a type_2 interaction which describes the *Mathematics_Chairman* component interacts with the *Mathematics_Lecturers* component.

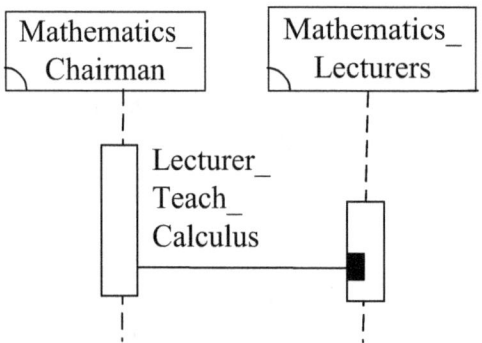

Figure 15-49   Interaction$_{g4}$

Interaction$_{h1}$ stands for the 1st interaction of the hth interaction flow diagram of the design view of *Kurdi University*, as shown in Figure 15-50. Interaction$_{h1}$ is a type_1 interaction which describes the *Student* actor interacts with the *University_President* component.

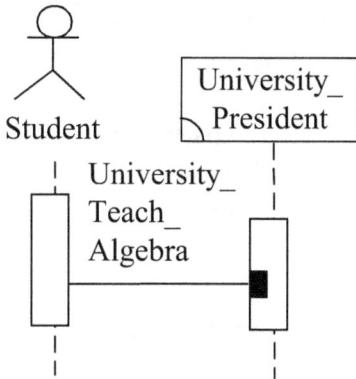

Figure 15-50   Interaction$_{h1}$

Interaction$_{h2}$ stands for the 2nd interaction of the hth interaction flow diagram of the design view of *Kurdi University*, as shown in Figure 15-51. Interaction$_{h2}$ is a type_2 interaction which describes the *University_President* component interacts with the *Science_Dean* component.

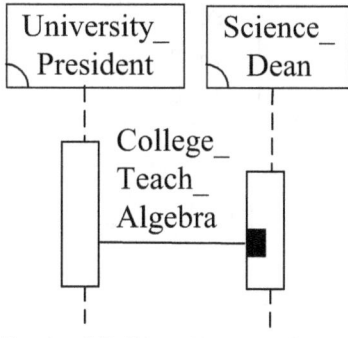

Figure 15-51   Interaction$_{h2}$

Interaction$_{h3}$ stands for the 3rd interaction of the hth interaction flow diagram of the design view of *Kurdi University*, as shown in Figure 15-52. Interaction$_{h3}$ is a type_2 interaction which describes the *Science_Dean* component interacts with the *Mathematics_Chairman* component.

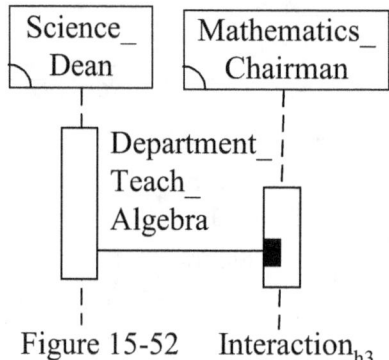

Figure 15-52   Interaction$_{h3}$

Interaction$_{h4}$ stands for the 4th interaction of the hth interaction flow diagram of the design view of *Kurdi University*, as shown in Figure 15-53. Interaction$_{h4}$ is a type_2 interaction which describes the *Mathematics_Chairman* component interacts with the *Mathematics_Lecturers* component.

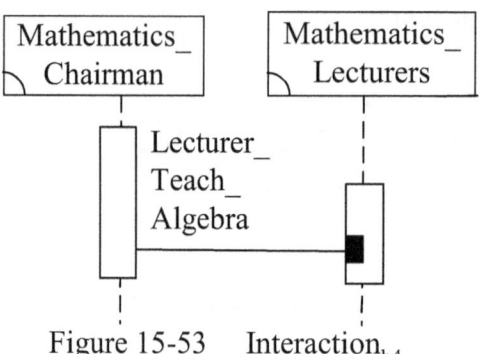

Figure 15-53   Interaction$_{h4}$

FixIFD$_g$ describes the recursion of the gth interaction flow diagram, i.e. *Study_Calculus_Course* behavior, of the design view of *Kurdi University*. FixIFD$_g$ is syntactically represented as "**fix**$(X_{07}$ = Interaction$_{g1}$●Interaction$_{g2}$●Interaction$_{g3}$●Interaction$_{g4}$●$X_{07})$", as shown in Figure 15-54.

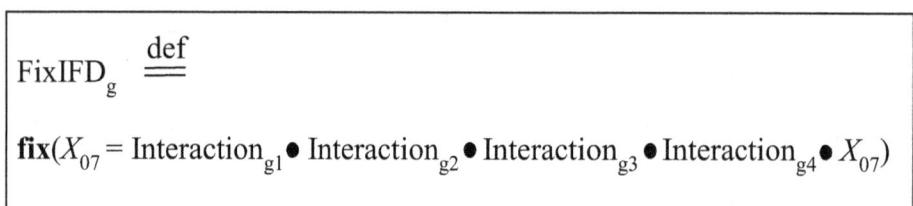

Figure 15-54   FixIFD$_g$

FixIFD$_h$ describes the recursion of the hth interaction flow diagram, i.e. *Study_Algebra_Course* behavior, of the design view of *Kurdi University*. FixIFD$_h$ is syntactically represented as "**fix**(X_{08} = Interaction$_{h1}$•Interaction$_{h2}$•Interaction$_{h3}$•Interaction$_{h4}$•X_{08})", as shown in Figure 15-55.

$$\text{FixIFD}_h \overset{\text{def}}{=\!=}$$

$$\textbf{fix}(X_{08} = \text{Interaction}_{h1} \bullet \text{Interaction}_{h2} \bullet \text{Interaction}_{h3} \bullet \text{Interaction}_{h4} \bullet X_{08})$$

Figure 15-55   FixIFD$_h$

Multi-queue SBC process of the design view of *Kurdi University* is syntactically represented as FixIFD$_g$‖FixIFD$_h$ which equals to "**fix**(X_{07}= Interaction$_{g1}$•Interaction$_{g2}$•Interaction$_{g3}$•Interaction$_{g4}$•X_{07})   ‖   **fix**(X_{08}= Interaction$_{h1}$•Interaction$_{h2}$•Interaction$_{h3}$•Interaction$_{h4}$•X_{08})", as shown in Figure 15-56.

$$\text{Kurdi University's Design View} \overset{\text{def}}{=\!=}$$

$$\textbf{fix}(X_{07} = \text{Interaction}_{g1} \bullet \text{Interaction}_{g2} \bullet \text{Interaction}_{g3} \bullet \text{Interaction}_{g4} \bullet X_{07}) \ \|$$
$$\textbf{fix}(X_{08} = \text{Interaction}_{h1} \bullet \text{Interaction}_{h2} \bullet \text{Interaction}_{h3} \bullet \text{Interaction}_{h4} \bullet X_{08})$$

Figure 15-56   Kurdi University's Design View

15-6 Multi-Queue SBC Process of the Structural Composition of the Kurdi University's Design View

Structural composition of the design view of *Kurdi University* means to compose the *Mathematics_Chairman* and *Mathematics_Lecturers* components into the *Mathematics_Department* component. That is, we will rename the *Mathematics_Chairman* component to the *Mathematics_Department* component; we also will rename the *Mathematics_Lecturers* component to the *Mathematics_Department* component, as shown in Figure 15-57.

[Mathematics_Department/Mathematics_Chairman,
Mathematics_Department/Mathematics_Lecturers]

Figure 15-57   Rename the *Mathematics_Chairman,
Mathematics_Lecturers* Components to the
Mathematics_Department Component

We draw the multi-queue SBC process algebra Backus-Naur Form tree of the structural composition of the design view of *Kurdi University* as shown in Figure 15-58.

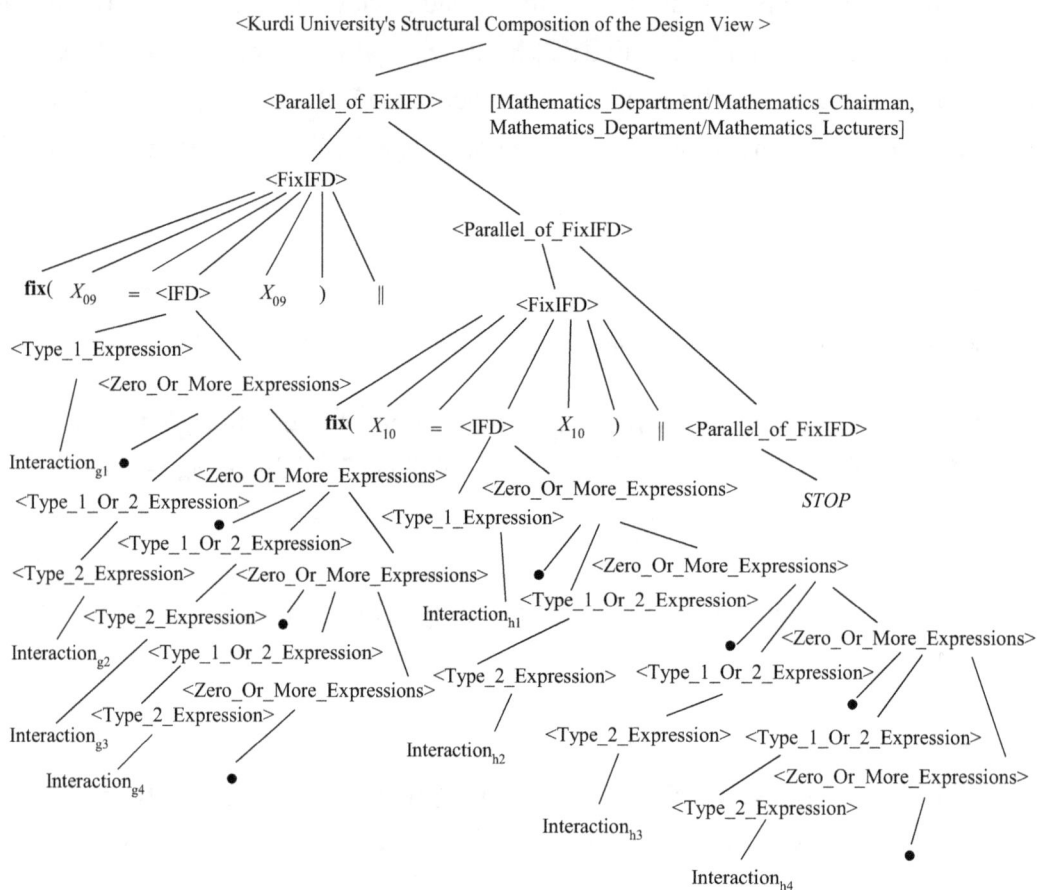

Figure 15-58   M-SBC-PA Backus-Naur Form Tree of the Structural Composition of the Design View of Kurdi University

Interaction$_{i1}$=Interaction$_{g1}$[Mathematics_Department/Mathematics_Chairman, Mathematics_Department/Mathematics_Lecturers]=Interaction$_{c1}$ stands for the 1st interaction of the ith interaction flow diagram of the structural composition of the design view of *Kurdi University*, as shown in Figure 15-59. Interaction$_{i1}$ is a type_1 interaction which describes the *Student* actor interacts with the *University_President* component.

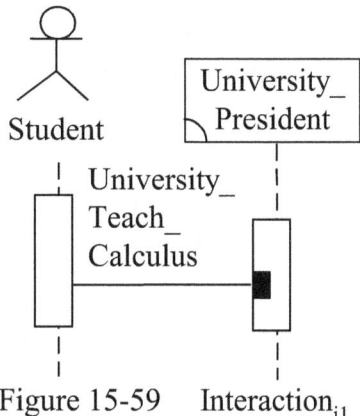

Figure 15-59   Interaction$_{i1}$

Interaction$_{i2}$=Interaction$_{g2}$[Mathematics_Department/Mathematics_Chairman, Mathematics_Department/Mathematics_Lecturers]=Interaction$_{c2}$ stands for the 2nd interaction of the ith interaction flow diagram of the structural composition of the design view of *Kurdi University*, as shown in Figure 15-60. Interaction$_{i2}$ is a type_2 interaction which describes the *University_President* component interacts with the *Science_Dean* component.

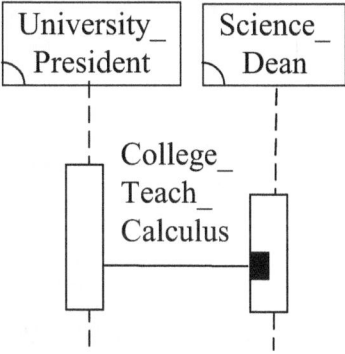

Figure 15-60   Interaction$_{i2}$

Interaction$_{i3}$=Interaction$_{g3}$[Mathematics_Department/Mathematics_Chairman, Mathematics_Department/Mathematics_Lecturers]=Interaction$_{c3}$ stands for the 3rd interaction of the ith interaction flow diagram of the structural composition of the design view of *Kurdi University*, as shown in Figure 15-61. Interaction$_{i3}$ is a type_2 interaction which describes the *Science_Dean* component interacts with the *Mathematics_Department* component.

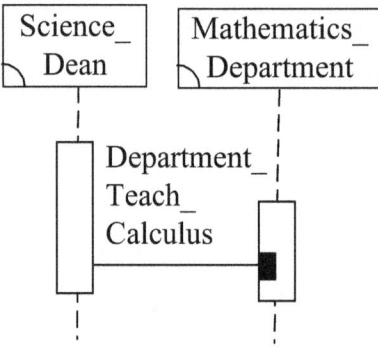

Figure 15-61   Interaction$_{i3}$

Interaction$_{i4}$=Interaction$_{g4}$[Mathematics_Department/Mathematics_Chairman, Mathematics_Department/Mathematics_Lecturers] stands for the 4th interaction of the ith interaction flow diagram of the structural composition of the design view of *Kurdi University*, as shown in Figure 15-62. Interaction$_{i4}$ is an internal interaction (i.e. λ) inside the *Mathematics_Department* component.

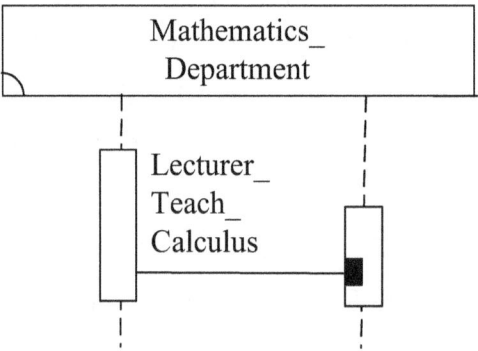

Figure 15-62   Interaction$_{i4}$

Interaction$_{j1}$=Interaction$_{h1}$[Mathematics_Department/Mathematics_Chairman, Mathematics_Department/Mathematics_Lecturers]=Interaction$_{d1}$ stands for the 1st interaction of the jth interaction flow diagram of the structural composition of the design view of *Kurdi University*, as shown in Figure 15-63. Interaction$_{j1}$ is a type_1 interaction which describes the *Student* actor interacts with the *University_President* component.

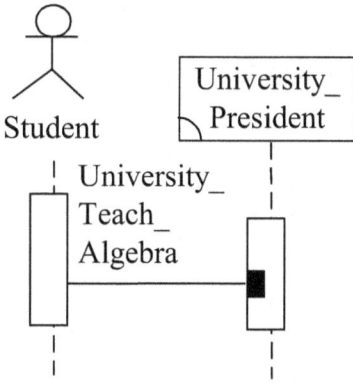

Figure 15-63   Interaction$_{j1}$

Interaction$_{j2}$=Interaction$_{h2}$[Mathematics_Department/Mathematics_Chairman, Mathematics_Department/Mathematics_Lecturers]=Interaction$_{d2}$ stands for the 2nd interaction of the jth interaction flow diagram of the structural composition of the design view of *Kurdi University*, as shown in Figure 15-64. Interaction$_{j2}$ is a type_2 interaction which describes the *University_President* component interacts with the *Science_Dean* component.

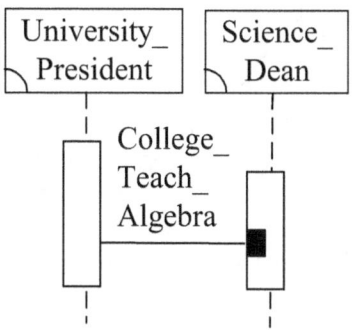

Figure 15-64   Interaction$_{j2}$

Interaction$_{j3}$=Interaction$_{h3}$[Mathematics_Department/Mathematics_Chairman, Mathematics_Department/Mathematics_Lecturers]=Interaction$_{d3}$ stands for the 3rd interaction of the jth interaction flow diagram of the structural composition of the design view of *Kurdi University*, as shown in Figure 15-65. Interaction$_{j3}$ is a type_2 interaction which describes the *Science_Dean* component interacts with the *Mathematics_Department* component.

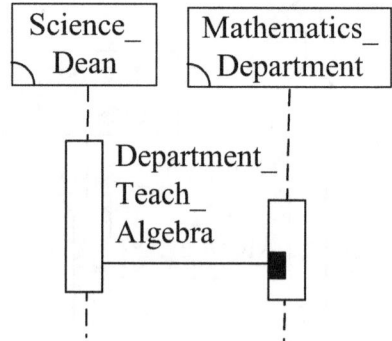

Figure 15-65   Interaction$_{j3}$

Interaction$_{j4}$=Interaction$_{h4}$[Mathematics_Department/Mathematics_Chairman, Mathematics_Department/Mathematics_Lecturers] stands for the 4th interaction of the jth interaction flow diagram of the structural composition of the design view of *Kurdi University*, as shown in Figure 15-66. Interaction$_{j4}$ is an internal interaction (i.e. λ) inside the *Mathematics_Department* component.

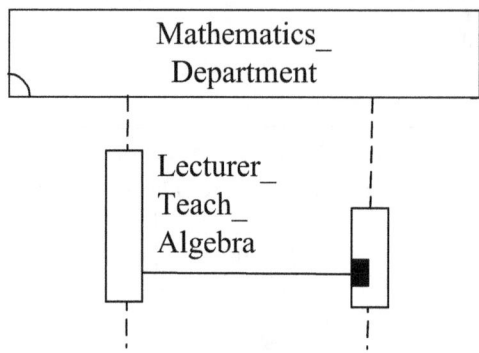

Figure 15-66   Interaction$_{j4}$

FixIFD$_i$ describes the recursion of the ith interaction flow diagram, i.e. *Study_Calculus_Course* behavior, of the structural composition of the design view of *Kurdi University*. FixIFD$_i$ is syntactically represented as "**fix**(X_{09} = Interaction$_{c1}$●Interaction$_{c2}$●Interaction$_{c3}$●λ●X_{09})" as shown in Figure 15-67.

$$\text{FixIFD}_i \stackrel{\text{def}}{=\!=}$$

$$\textbf{fix}(X_{09} = \text{Interaction}_{c1} \bullet \text{Interaction}_{c2} \bullet \text{Interaction}_{c3} \bullet \lambda \bullet X_{09})$$

Figure 15-67   FixIFD$_i$

FixIFD$_j$ describes the recursion of the jth interaction flow diagram, i.e. *Study_Algebra_Course* behavior, of the structural composition of the design view of *Kurdi University*. FixIFD$_j$ is syntactically represented as "$\textbf{fix}(X_{10} = \text{Interaction}_{d1} \bullet \text{Interaction}_{d2} \bullet \text{Interaction}_{d3} \bullet \lambda \bullet X_{10})$" as shown in Figure 15-68.

$$\text{FixIFD}_j \stackrel{\text{def}}{=\!=}$$

$$\textbf{fix}(X_{10} = \text{Interaction}_{d1} \bullet \text{Interaction}_{d2} \bullet \text{Interaction}_{d3} \bullet \lambda \bullet X_{10})$$

Figure 15-68   FixIFD$_j$

Multi-queue SBC process of the structural composition of the design view of *Kurdi University* is syntactically represented as FixIFD$_i$||FixIFD$_j$ which equals to "$\textbf{fix}(X_{09} = \text{Interaction}_{c1} \bullet \text{Interaction}_{c2} \bullet \text{Interaction}_{c3} \bullet \lambda \bullet X_{09})$ || $\textbf{fix}(X_{10} = \text{Interaction}_{d1} \bullet \text{Interaction}_{d2} \bullet \text{Interaction}_{d3} \bullet \lambda \bullet X_{10})$" as shown in Figure 15-69.

Kurdi University's Structural Composition of the Design View $\stackrel{\text{def}}{=\!=}$

$$\textbf{fix}(X_{09} = \text{Interaction}_{c1} \bullet \text{Interaction}_{c2} \bullet \text{Interaction}_{c3} \bullet \lambda \bullet X_{09}) \; || $$
$$\textbf{fix}(X_{10} = \text{Interaction}_{d1} \bullet \text{Interaction}_{d2} \bullet \text{Interaction}_{d3} \bullet \lambda \bullet X_{10})$$

Figure 15-69   Kurdi University's Structural Composition
of the Design View

15-7 Observation Congruence of "the Analysis View" and "the Structural Composition of the Design View" of Kurdi University

We syntactically represent multi-queue SBC process P_{01} as FixIFD$_c$‖FixIFD$_d$ which equals to "$\textbf{fix}(X_{03}=\text{Interaction}_{c1}\bullet\text{Interaction}_{c2}\bullet\text{Interaction}_{c3}\bullet X_{03})$ ‖ $\textbf{fix}(X_{04}=\text{Interaction}_{d1}\bullet\text{Interaction}_{d2}\bullet\text{Interaction}_{d3}\bullet X_{04})$." The transition graph, as shown in Figure 15-70, demonstrates the semantics of process P_{01}.

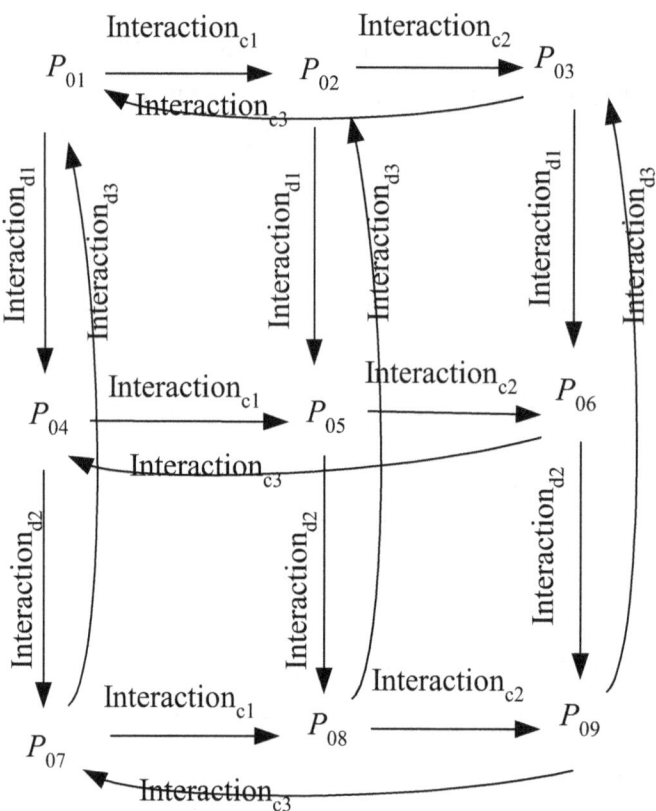

Figure 15-70 Transition Graph Demonstrates the Semantics of Process P_{01}

We syntactically represent process Q_{01} as FixIFD$_i$‖FixIFD$_j$ which equals to "$\textbf{fix}(X_{09} = \text{Interaction}_{c1}\bullet\text{Interaction}_{c2}\bullet\text{Interaction}_{c3}\bullet\lambda\bullet X_{09})$ ‖ $\textbf{fix}(X_{10} = \text{Interaction}_{d1}\bullet\text{Interaction}_{d2}\bullet\text{Interaction}_{d3}\bullet\lambda\bullet X_{10})$." The transition graph, as shown in Figure 15-71, demonstrates the semantics of process Q_{01}.

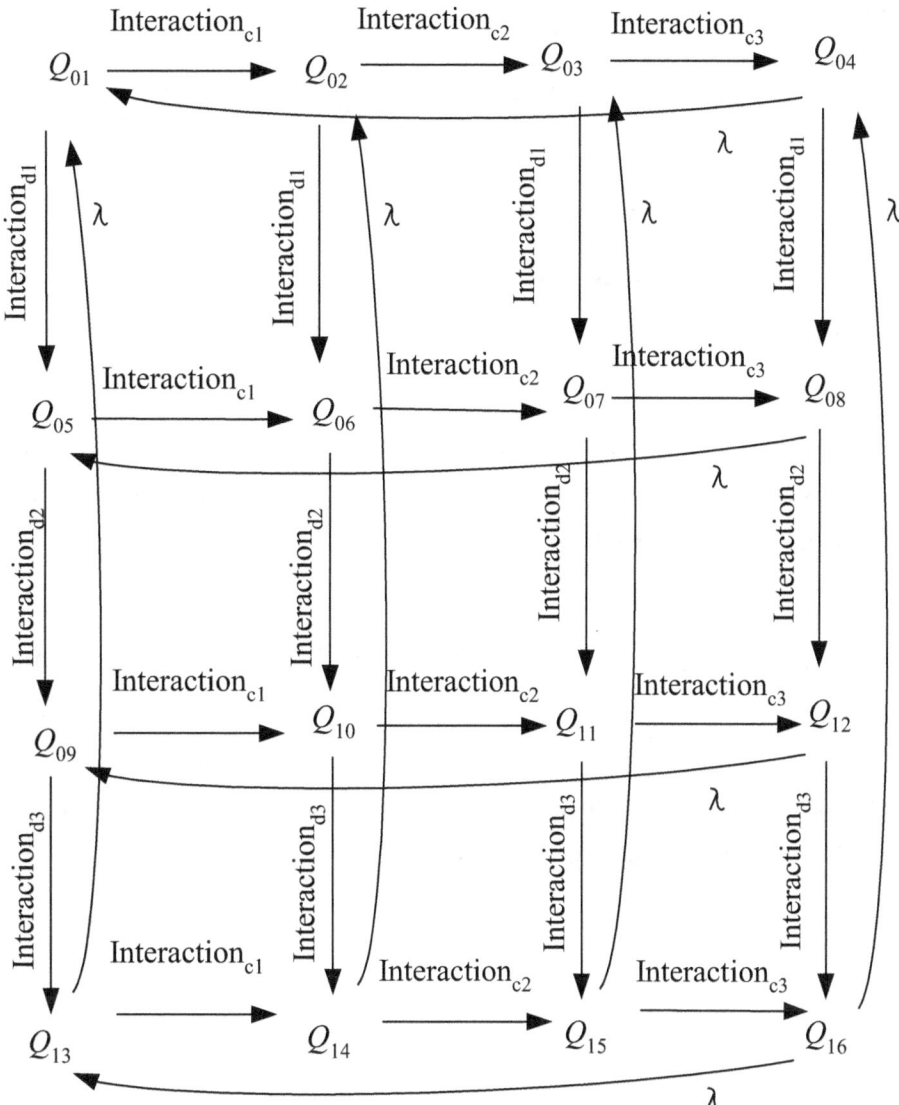

Figure 15-71   Transition Graph Demonstrates the Semantics of Process Q_{01}

We can easily verify that $S = \{(P_{01}, Q_{01}), (P_{01}, Q_{04}), (P_{01}, Q_{13}), (P_{01}, Q_{16}), (P_{02}, Q_{02}), (P_{02}, Q_{14}), (P_{03}, Q_{03}), (P_{03}, Q_{15}), (P_{04}, Q_{05}), (P_{04}, Q_{08}), (P_{05}, Q_{06}), (P_{06}, Q_{07}), (P_{07}, Q_{09}), (P_{07}, Q_{12}), (P_{08}, Q_{10}), (P_{09}, Q_{11})\}$ is a bisimulation.

Using the S bisimulation, we then are able to verify that P_{01} and Q_{01} are observation congruent because (1) $P_{01} \xrightarrow{\text{Interaction}_{c1}} P_{02}$, then we have Q_{02} that Q_{01} $\xRightarrow{\text{Interaction}_{c1}} Q_{02}$ and $P_{02} \overset{\approx}{\sim} Q_{02}$, and (2) $P_{01} \xrightarrow{\text{Interaction}_{d1}} P_{04}$, then we have Q_{05} that

$Q_{01} \xLongrightarrow{\text{Interaction}_{d1}} Q_{05}$ and $P_{04} \approx Q_{05}$, and (3) $Q_{01} \xrightarrow{\text{Interaction}_{c1}} Q_{02}$, then we have P_{02}

that $P_{01} \xLongrightarrow{\text{Interaction}_{c1}} P_{02}$ and $P_{02} \approx Q_{02}$, and (4) $Q_{01} \xrightarrow{\text{Interaction}_{d1}} Q_{05}$, then we have

P_{04} that $P_{01} \xLongrightarrow{\text{Interaction}_{d1}} P_{04}$ and $P_{04} \approx Q_{05}$.

$P_{01} = Q_{01}$ means that "the analysis view of *Kurdi University*" and "the structural composition of the design view of *Kurdi University*" are observation congruent.

That is, we have demonstrated that the analysis view of *Kurdi University* is one level up structural composition (with observation congruence verification) of the design view of *Kurdi University*.

15-8 Multi-Queue SBC Process of the Kurdi University's Implementation View

We draw the Architecture Hierarchy Diagram (AHD) of the design view of *Kurdi University* as shown in Figure 15-72.

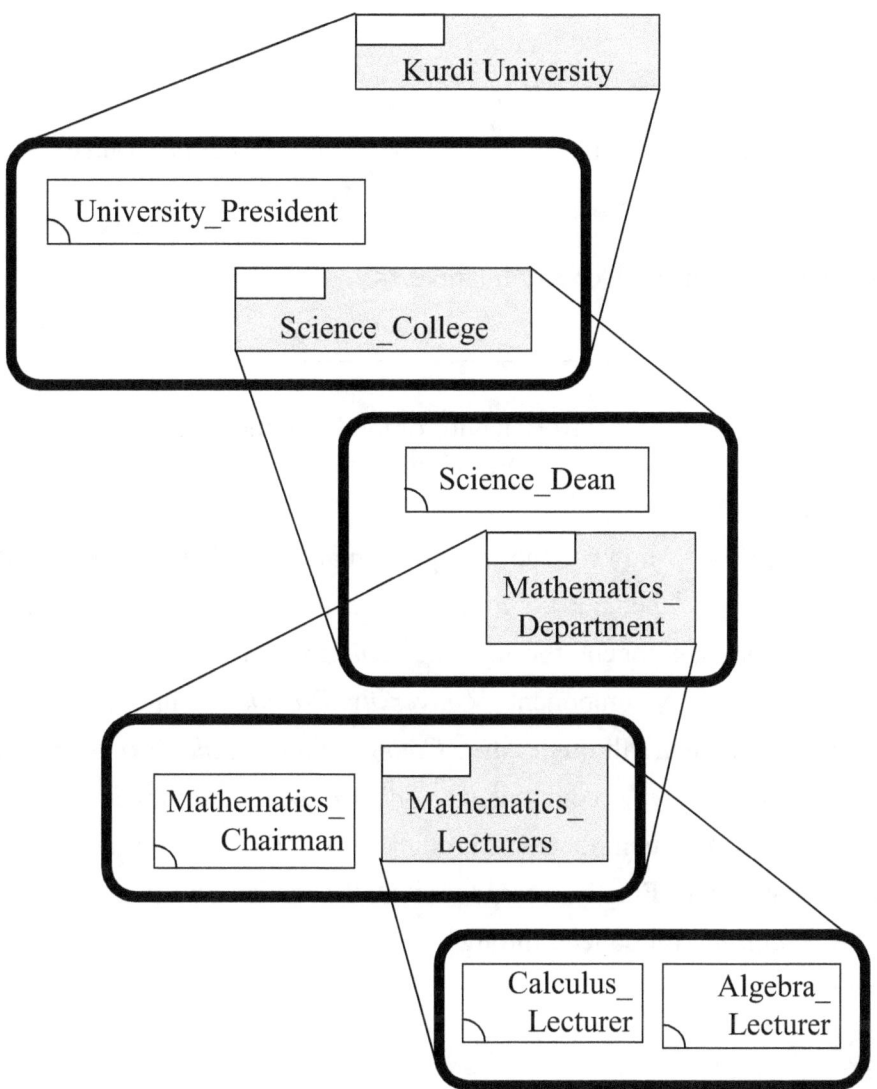

Figure 15-72   AHD of the Implementation View of *Kurdi University*

The overall behavior of the implementation view of *Kurdi University* includes two behaviors: *Study_Calculus_Course* and *Study_Algebra_Course* as shown in Figure 15-73. Each of them is described by an individual IFD.

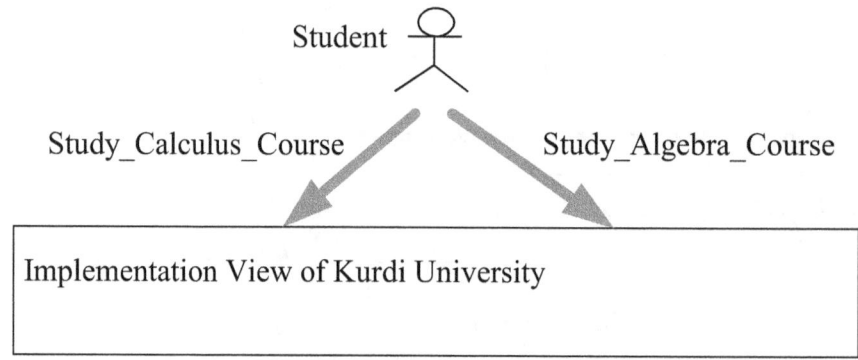

Figure 15-73   Overall Behavior of the Implementation View
of Kurdi University

Figure 15-74 shows the implementation view's IFD of the *Study_Calculus_Course* behavior. First, actor *Student* interacts with the *University_President* component through the *University_Teach_Calculus* operation call interaction. Next, component *University_President* interacts with the *Science_Dean* component through the *College_Teach_Calculus* operation call interaction. Continuingly, component *Science_Dean* interacts with the *Mathematics_Chairman* component through the *Department_Teach_Calculus* operation call interaction. Finally, component *Mathematics_Chairman* interacts with the *Calculus_Lecturer* component through the *Lecturer_Teach_Calculus* operation call interaction.

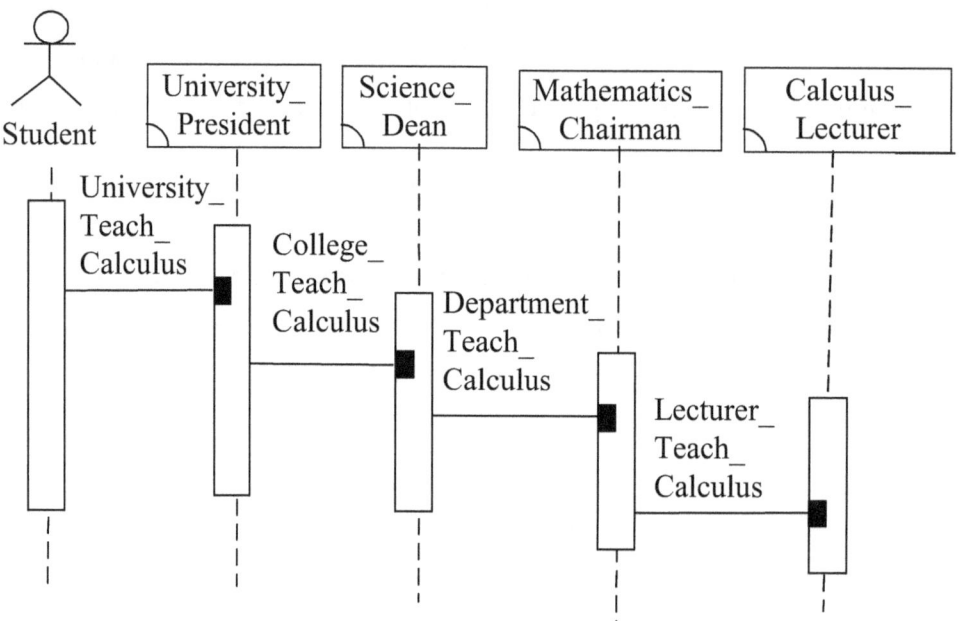

Figure 15-74   Implementation View's IFD of the *Study_Calculus_Course* Behavior

Figure 15-75 shows the implementation view's IFD of the *Study_Algebra_Course* behavior. First, actor *Student* interacts with the *University_President* component through the *University_Teach_Algebra* operation call interaction. Next, component *University_President* interacts with the *Science_Dean* component through the *College_Teach_Algebra* operation call interaction. Continuingly, component *Science_Dean* interacts with the *Mathematics_Chairman* component through the *Department_Teach_Algebra* operation call interaction. Finally, component *Mathematics_Chairman* interacts with the *Algebra_Lecturer* component through the *Lecturer_Teach_Algebra* operation call interaction.

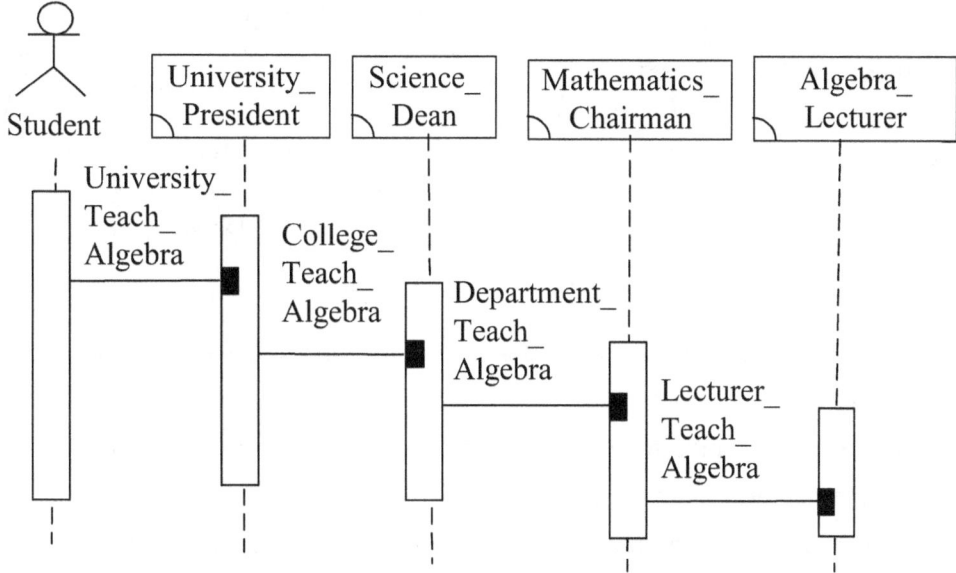

Figure 15-75   Implementation View's IFD of the *Study_Algebra_Course* Behavior

We draw the multi-queue SBC process algebra Backus-Naur Form tree of the implementation view of *Kurdi University* as shown in Figure 15-76.

184

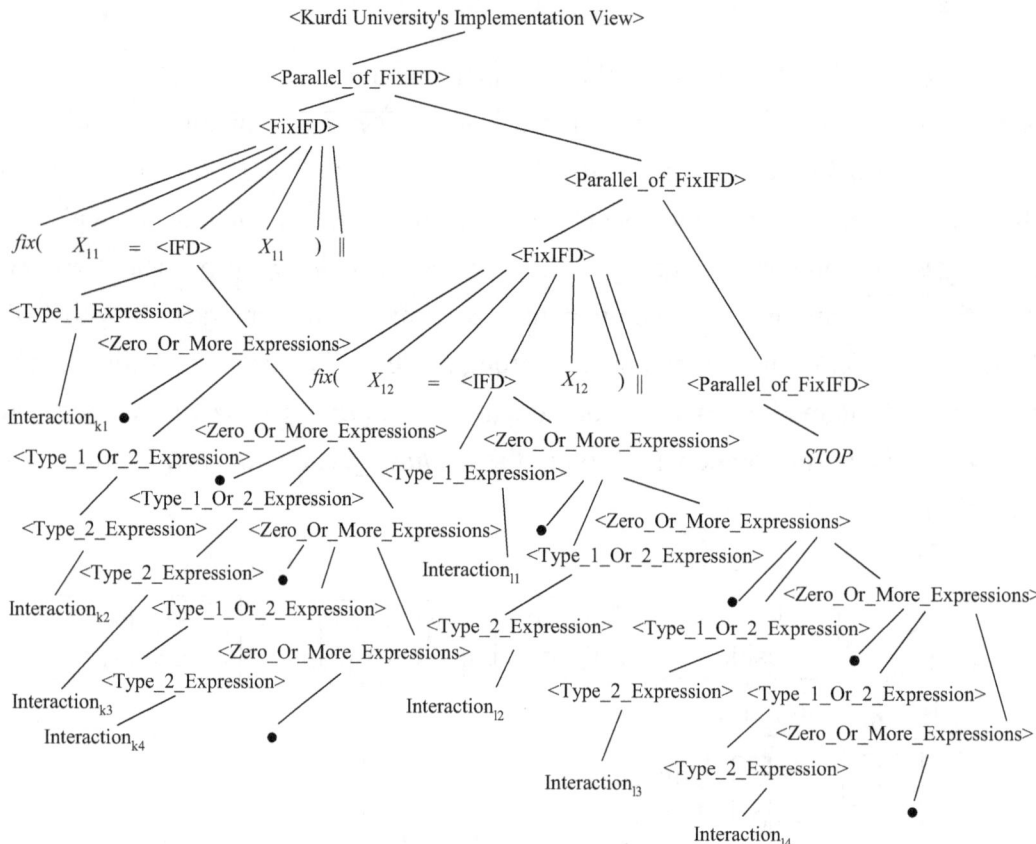

Figure 15-76   M-SBC-PA Backus-Naur Form Tree of the Implementation View of Kurdi University

Interaction$_{k1}$ stands for the 1st interaction of the kth interaction flow diagram of the implementation view of *Kurdi University*, as shown in Figure 15-77. Interaction$_{k1}$ is a type_1 interaction which describes the *Student* actor interacts with the *University_President* component.

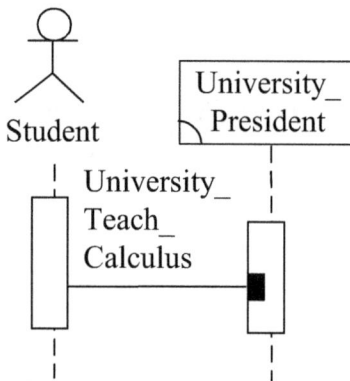

Figure 15-77   Interaction$_{k1}$

Interaction$_{k2}$ stands for the 2nd interaction of the kth interaction flow diagram of the implementation view of *Kurdi University*, as shown in Figure 15-78. Interaction$_{k2}$ is a type_2 interaction which describes the *University_President* component interacts with the *Science_Dean* component.

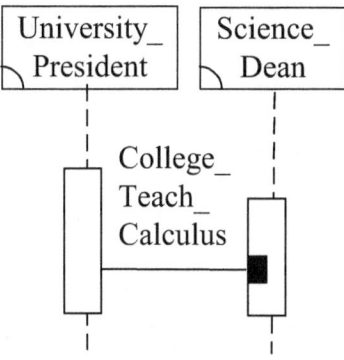

Figure 15-78   Interaction$_{k2}$

Interaction$_{k3}$ stands for the 3rd interaction of the kth interaction flow diagram of the implementation view of *Kurdi University*, as shown in Figure 15-79. Interaction$_{k3}$ is a type_2 interaction which describes the *Science_Dean* component interacts with the *Mathematics_Chairman* component.

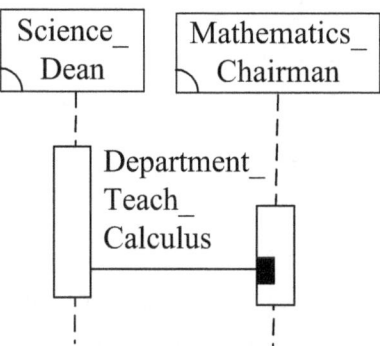

Figure 15-79   Interaction$_{k3}$

Interaction$_{k4}$ stands for the 4th interaction of the kth interaction flow diagram of the implementation view of *Kurdi University*, as shown in Figure 15-80. Interaction$_{k4}$ is a type_2 interaction which describes the *Mathematics_Chairman* component interacts with the *Calculus_Lecturer* component.

186

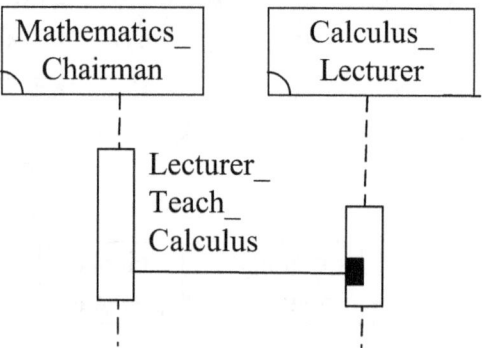

Figure 15-80   Interaction$_{k4}$

Interaction$_{l1}$ stands for the 1st interaction of the lth interaction flow diagram of the implementation view of *Kurdi University*, as shown in Figure 15-81. Interaction$_{l1}$ is a type_1 interaction which describes the *Student* actor interacts with the *University_President* component.

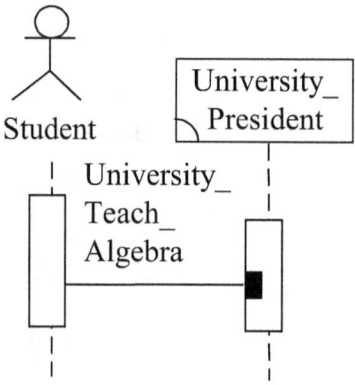

Figure 14-81   Interaction$_{l1}$

Interaction$_{l2}$ stands for the 2nd interaction of the lth interaction flow diagram of the implementation view of *Kurdi University*, as shown in Figure 15-82. Interaction$_{l2}$ is a type_2 interaction which describes the *University_President* component interacts with the *Science_Dean* component.

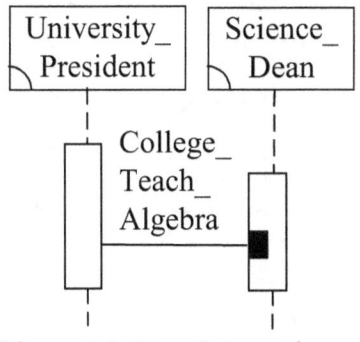

Figure 15-82   Interaction$_{12}$

Interaction$_{13}$ stands for the 3rd interaction of the lth interaction flow diagram of the implementation view of *Kurdi University*, as shown in Figure 15-83. Interaction$_{13}$ is a type_2 interaction which describes the *Science_Dean* component interacts with the *Mathematics_Chairman* component.

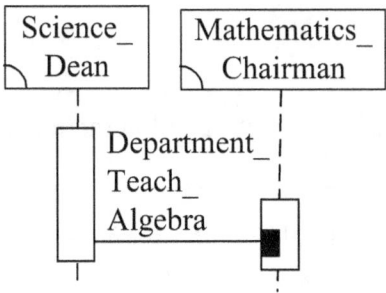

Figure 15-83   Interaction$_{13}$

Interaction$_{14}$ stands for the 4th interaction of the lth interaction flow diagram of the implementation view of *Kurdi University*, as shown in Figure 15-84. Interaction$_{14}$ is a type_2 interaction which describes the *Mathematics_Chairman* component interacts with the *Algebra_Lecturer* component.

188

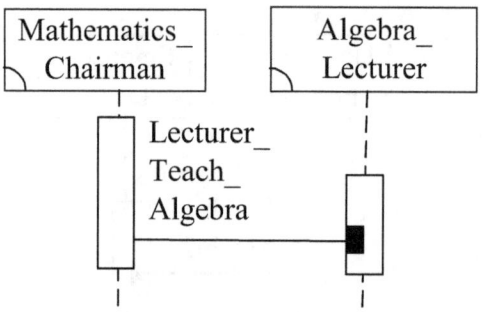

Figure 15-84   Interaction$_{l4}$

FixIFD$_k$ describes the recursion of the kth interaction flow diagram, i.e. *Study_Calculus_Course* behavior, of the implementation view of *Kurdi University*. FixIFD$_k$ is syntactically represented as "**fix**(X_{11} = Interaction$_{k1}$●Interaction$_{k2}$●Interaction$_{k3}$●Interaction$_{k4}$●X_{11})", as shown in Figure 15-85.

$$
\text{FixIFD}_k \overset{\text{def}}{=\!=}
$$

$$
\mathbf{fix}(X_{11} = \text{Interaction}_{k1} \bullet \text{Interaction}_{k2} \bullet \text{Interaction}_{k3} \bullet \text{Interaction}_{k4} \bullet X_{11})
$$

Figure 15-85   FixIFD$_k$

FixIFD$_l$ describes the recursion of the lth interaction flow diagram, i.e. *Study_Algebra_Course* behavior, of the implementation view of *Kurdi University*. FixIFD$_l$ is syntactically represented as "**fix**(X_{12} = Interaction$_{l1}$●Interaction$_{l2}$●Interaction$_{l3}$●Interaction$_{l4}$●X_{12})", as shown in Figure 15-86.

$$
\text{FixIFD}_l \overset{\text{def}}{=\!=}
$$

$$
\mathbf{fix}(X_{12} = \text{Interaction}_{l1} \bullet \text{Interaction}_{l2} \bullet \text{Interaction}_{l3} \bullet \text{Interaction}_{l4} \bullet X_{12})
$$

Figure 15-86   FixIFD$_l$

Multi-queue SBC process of the implementation view of *Kurdi University* is syntactically represented as $\text{FixIFD}_k \| \text{FixIFD}_l$ which equals to "**fix**(X_{11} = Interaction$_{k1}$●Interaction$_{k2}$●Interaction$_{k3}$●Interaction$_{k4}$●X_{11}) ‖ **fix**(X_{12} = Interaction$_{l1}$●Interaction$_{l2}$●Interaction$_{l3}$●Interaction$_{l4}$●X_{12})", as shown in Figure 15-87.

Kurdi University's Implementation View $\overset{\text{def}}{=\!=}$

fix(X_{11}=Interaction$_{k1}$ ● Interaction$_{k2}$ ● Interaction$_{k3}$●Interaction$_{k4}$ ●X_{11}) ‖
fix(X_{12}=Interaction$_{l1}$ ● Interaction$_{l2}$ ● Interaction$_{l3}$ ● Interaction$_{l4}$ ●X_{12})

Figure 15-87   Kurdi University's Implementation View

15-9 Multi-Queue SBC Process of the Structural Composition of the Kurdi University's Implementation View

Structural composition of the implementation view of *Kurdi University* means to compose the *Calculus_Lecturer* and *Algebra_Lecturer* components into the *Mathematics_Lecturers* component. That is, we will rename the *Calculus_Lecturer* component to the *Mathematics_Lecturers* component; we also will rename the *Algebra_Lecturer* component to the *Mathematics_Lecturers* component, as shown in Figure 15-88.

[Mathematics_Lecturers/Calculus_Lecturer,
Mathematics_Lecturers/Algebra_Lecturer]

Figure 15-88   Rename the *Calculus_Lecturer*, *Algebra_Lecturer*
Components to the *Mathematics_Lecturers* Component

We draw the Backus-Naur Form tree of the structural composition of the implementation view of *Kurdi University* as shown in Figure 15-89.

190

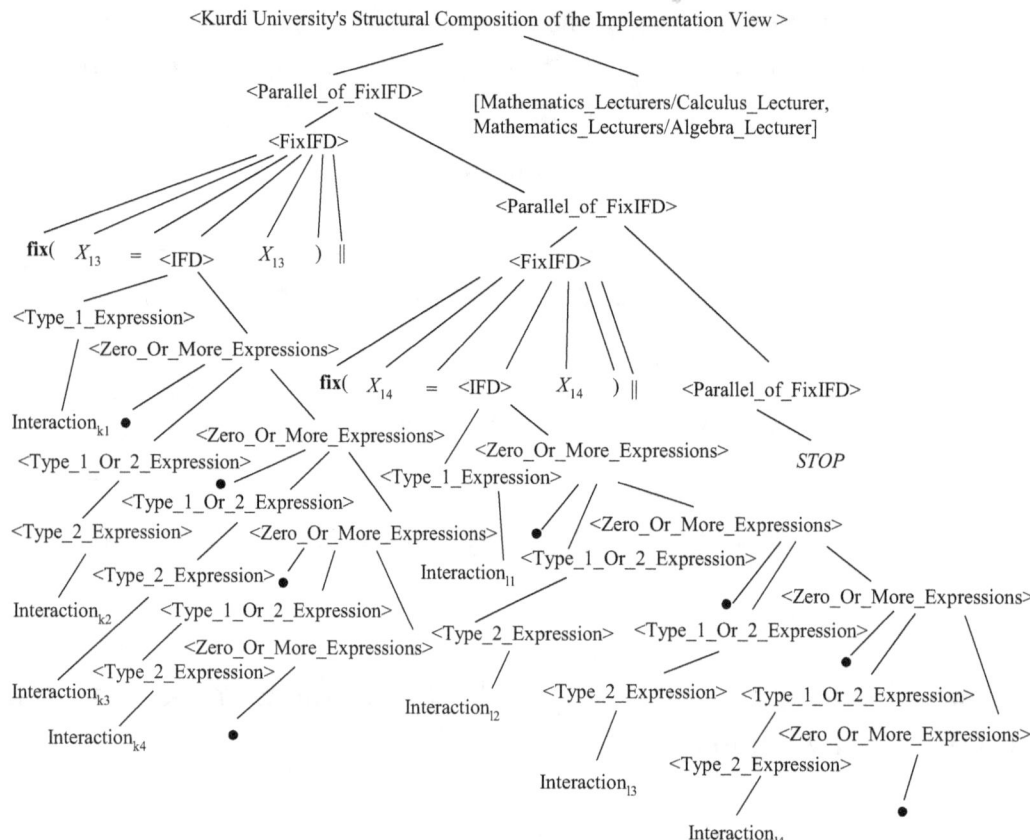

Figure 15-89   M-SBC-PA Backus-Naur Form Tree of the Structural Composition
of the Implementation View of Kurdi University

Interaction$_{m1}$=Interaction$_{k1}$[Mathematics_Lecturers/Calculus_Lecturer,Mathe
matics_Lecturers/Algebra_Lecturer]=Interaction$_{g1}$ stands for the 1st interaction of the
mth interaction flow diagram of the structural composition of the implementation
view of *Kurdi University*, as shown in Figure 15-90. Interaction$_{m1}$ is a type_1
interaction which describes the *Student* actor interacts with the *University_President*
component.

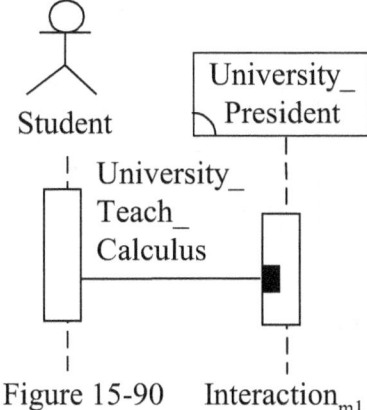

Figure 15-90   Interaction$_{m1}$

Interaction$_{m2}$=Interaction$_{k2}$[Mathematics_Lecturers/Calculus_Lecturer,Mathematics_Lecturers/Algebra_Lecturer]=Interaction$_{g2}$ stands for the 2nd interaction of the mth interaction flow diagram of the structural composition of the implementation view of *Kurdi University*, as shown in Figure 15-91. Interaction$_{m2}$ is a type_2 interaction which describes the *University_President* component interacts with the *Science_Dean* component.

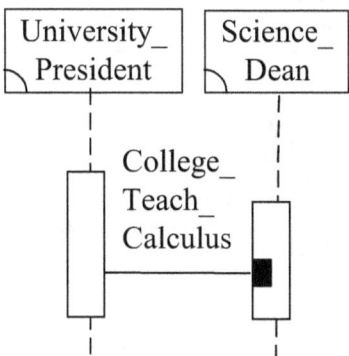

Figure 15-91   Interaction$_{m2}$

Interaction$_{m3}$=Interaction$_{k3}$[Mathematics_Lecturers/Calculus_Lecturer,Mathematics_Lecturers/Algebra_Lecturer]=Interaction$_{g3}$ stands for the 3rd interaction of the mth interaction flow diagram of the structural composition of the implementation view of *Kurdi University*, as shown in Figure 15-92. Interaction$_{m3}$ is a type_2 interaction which describes the *Science_Dean* component interacts with the *Mathematics_Chairman* component.

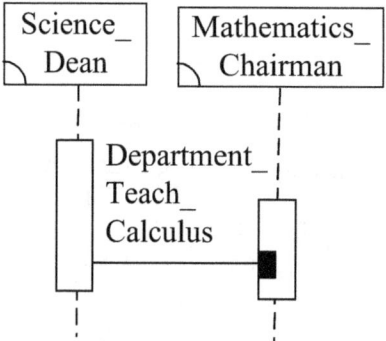

Figure 15-92   Interaction$_{m3}$

Interaction$_{m4}$=Interaction$_{k4}$[Mathematics_Lecturers/Calculus_Lecturer,Mathematics_Lecturers/Algebra_Lecturer]=Interaction$_{g4}$ stands for the 4th interaction of the mth interaction flow diagram of the structural composition of the implementation view of *Kurdi University*, as shown in Figure 15-93. Interaction$_{m4}$ is a type_2 interaction which describes the *Mathematics_Chairman* component interacts with the *Mathematics_Lecturers* component.

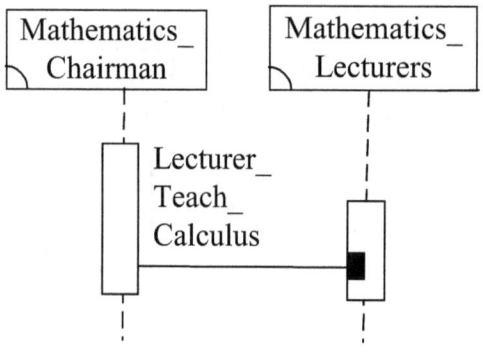

Figure 15-93   Interaction$_{m4}$

Interaction$_{n1}$=Interaction$_{l1}$[Mathematics_Lecturers/Calculus_Lecturer,Mathematics_Lecturers/Algebra_Lecturer]=Interaction$_{h1}$ stands for the 1st interaction of the nth interaction flow diagram of the structural composition of the implementation view of *Kurdi University*, as shown in Figure 15-94. Interaction$_{n1}$ is a type_1 interaction which describes the *Student* actor interacts with the *University_President* component.

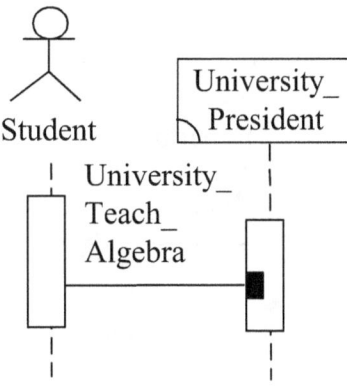

Figure 15-94   Interaction$_{n1}$

Interaction$_{n2}$=Interaction$_{l2}$[Mathematics_Lecturers/Calculus_Lecturer,Mathem atics_Lecturers/Algebra_Lecturer]=Interaction$_{h2}$ stands for the 2nd interaction of the nth interaction flow diagram of the structural composition of the implementation view of *Kurdi University*, as shown in Figure 15-95. Interaction$_{n2}$ is a type_2 interaction which describes the *University_President* component interacts with the *Science_Dean* component.

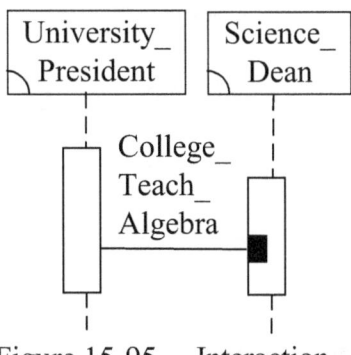

Figure 15-95   Interaction$_{n2}$

Interaction$_{n3}$=Interaction$_{l3}$[Mathematics_Lecturers/Calculus_Lecturer,Mathem atics_Lecturers/Algebra_Lecturer]=Interaction$_{h3}$ stands for the 3rd interaction of the nth interaction flow diagram of the structural composition of the implementation view of *Kurdi University*, as shown in Figure 15-96. Interaction$_{n3}$ is a type_2 interaction which describes the *Science_Dean* component interacts with the *Mathematics_Chairman* component.

194

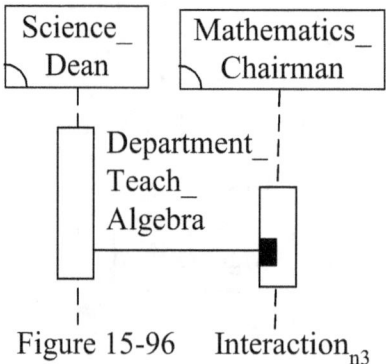

Figure 15-96   Interaction$_{n3}$

Interaction$_{n4}$=Interaction$_{l4}$[Mathematics_Lecturers/Calculus_Lecturer,Mathem atics_Lecturers/Algebra_Lecturer]=Interaction$_{h4}$ stands for the 4th interaction of the nth interaction flow diagram of the structural composition of the implementation view of *Kurdi University*, as shown in Figure 15-97. Interaction$_{n4}$ is a type_2 interaction which describes the *Mathematics_Chairman* component interacts with the *Mathematics_Lecturers* component.

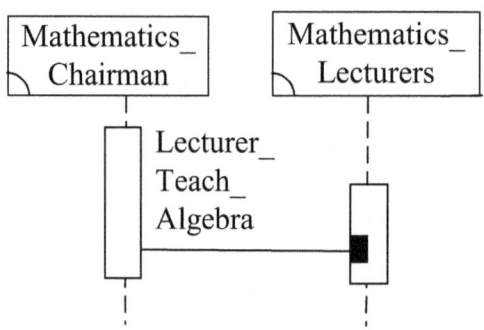

Figure 15-97   Interaction$_{n4}$

FixIFD$_{m}$ describes the recursion of the mth interaction flow diagram, i.e. *Study_Calculus_Course* behavior, of the structural composition of the implementation view of *Kurdi University*. FixIFD$_{m}$ is syntactically represented as "**fix**(X_{13} = Interaction$_{g1}$●Interaction$_{g2}$●Interaction$_{g3}$●Interaction$_{g4}$●X_{13})", as shown in Figure 15-98.

$$\text{FixIFD}_m \overset{\text{def}}{=\!=}$$

$$\textbf{fix}(X_{13} = \text{Interaction}_{g1} \bullet \text{Interaction}_{g2} \bullet \text{Interaction}_{g3} \bullet \text{Interaction}_{g4} \bullet X_{13})$$

Figure 15-98 FixIFD_m

FixIFD_n describes the recursion of the nth interaction flow diagram, i.e. *Study_Algebra_Course* behavior, of the structural composition of the implementation view of *Kurdi University*. FixIFD_n is syntactically represented as "$\textbf{fix}(X_{14} = \text{Interaction}_{h1} \bullet \text{Interaction}_{h2} \bullet \text{Interaction}_{h3} \bullet \text{Interaction}_{h4} \bullet X_{14})$", as shown in Figure 15-99.

$$\text{FixIFD}_n \overset{\text{def}}{=\!=}$$

$$\textbf{fix}(X_{14} = \text{Interaction}_{h1} \bullet \text{Interaction}_{h2} \bullet \text{Interaction}_{h3} \bullet \text{Interaction}_{h4} \bullet X_{14})$$

Figure 15-99 FixIFD_n

Multi-queue SBC process of the structural composition of the implementation view of *Kurdi University* is syntactically represented as $FixIFD_m \| FixIFD_n$ which equals to "$\mathbf{fix}(X_{13}=$ $Interaction_{g1} \bullet Interaction_{g2} \bullet Interaction_{g3} \bullet Interaction_{g4} \bullet X_{13})$ $\|$ $\mathbf{fix}(X_{14} = Interaction_{h1} \bullet Interaction_{h2} \bullet Interaction_{h3} \bullet Interaction_{h4} \bullet X_{14})$", as shown in Figure 15-100.

Kurdi University's Structural Composition of the Implementation View $\overset{def}{=\!=}$

$$\mathbf{fix}(X_{13}=Interaction_{g1} \bullet Interaction_{g2} \bullet Interaction_{g3} \bullet Interaction_{g4} \bullet X_{13}) \|$$
$$\mathbf{fix}(X_{14}=Interaction_{h1} \bullet Interaction_{h2} \bullet Interaction_{h3} \bullet Interaction_{h4} \bullet X_{14})$$

Figure 15-100   Kurdi University's Structural Composition
of the Implementation View

15-10 Observation Congruence of "the Design View" and "the Structural Composition of the Implementation View" of Kurdi University

We syntactically represent multi-queue SBC process P_{01} as $FixIFD_g \| FixIFD_h$ which equals to "$\mathbf{fix}(X_{07}= Interaction_{g1} \bullet Interaction_{g2} \bullet Interaction_{g3} \bullet Interaction_{g4} \bullet X_{07})$ $\|$ $\mathbf{fix}(X_{08}= Interaction_{h1} \bullet Interaction_{h2} \bullet Interaction_{h3} \bullet Interaction_{h4} \bullet X_{08})$." The transition graph, as shown in Figure 15-101, demonstrates the semantics of process P_{01}.

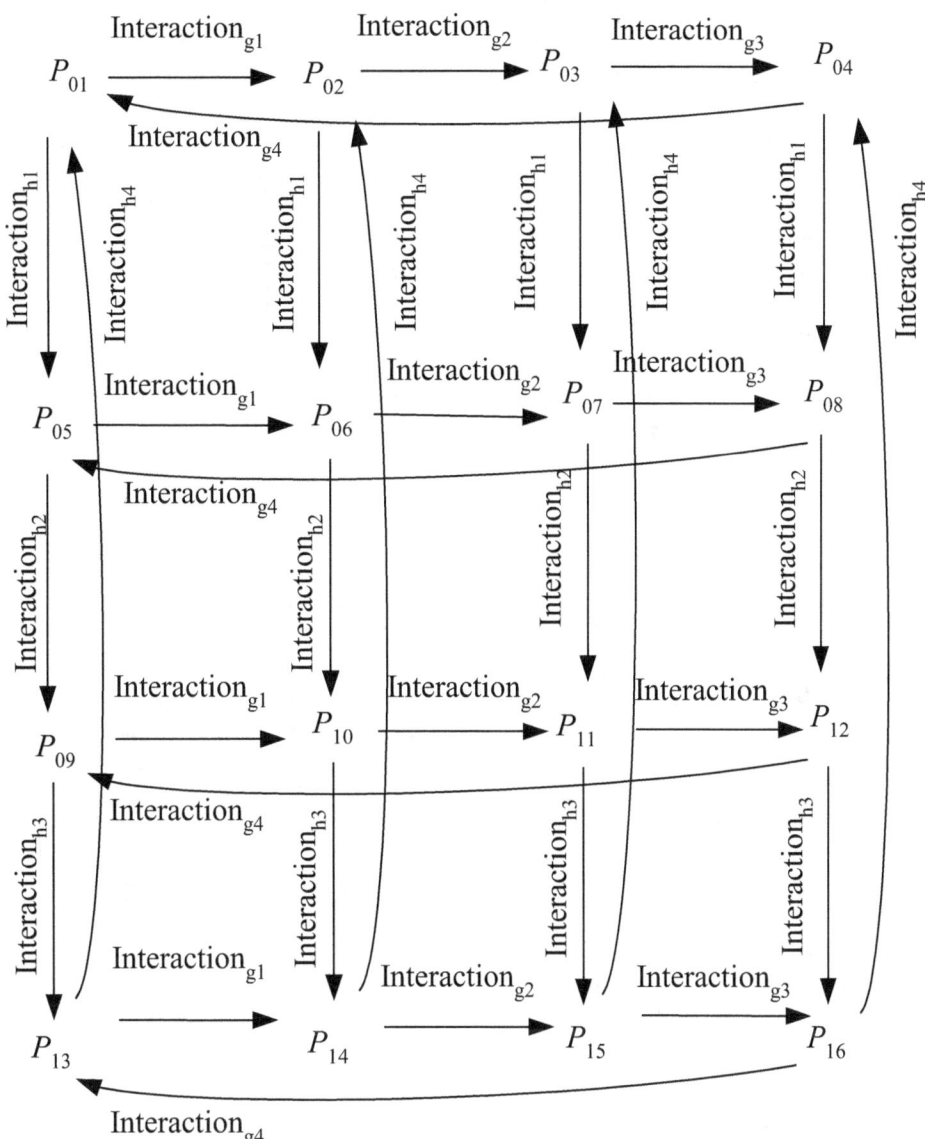

Figure 15-101　Transition Graph Demonstrates the Semantics of Process P_{01}

We syntactically represent multi-queue SBC process Q_{01} as $FixIFD_m \| FixIFD_n$ which equals to "**fix**(X_{13}= Interaction$_{g1}$•Interaction$_{g2}$•Interaction$_{g3}$•Interaction$_{g4}$•X_{13}) $\|$ **fix**(X_{14}= Interaction$_{h1}$•Interaction$_{h2}$•Interaction$_{h3}$•Interaction$_{h4}$•X_{14})." The transition graph, as shown in Figure 15-102, demonstrates the semantics of process Q_{01}.

198

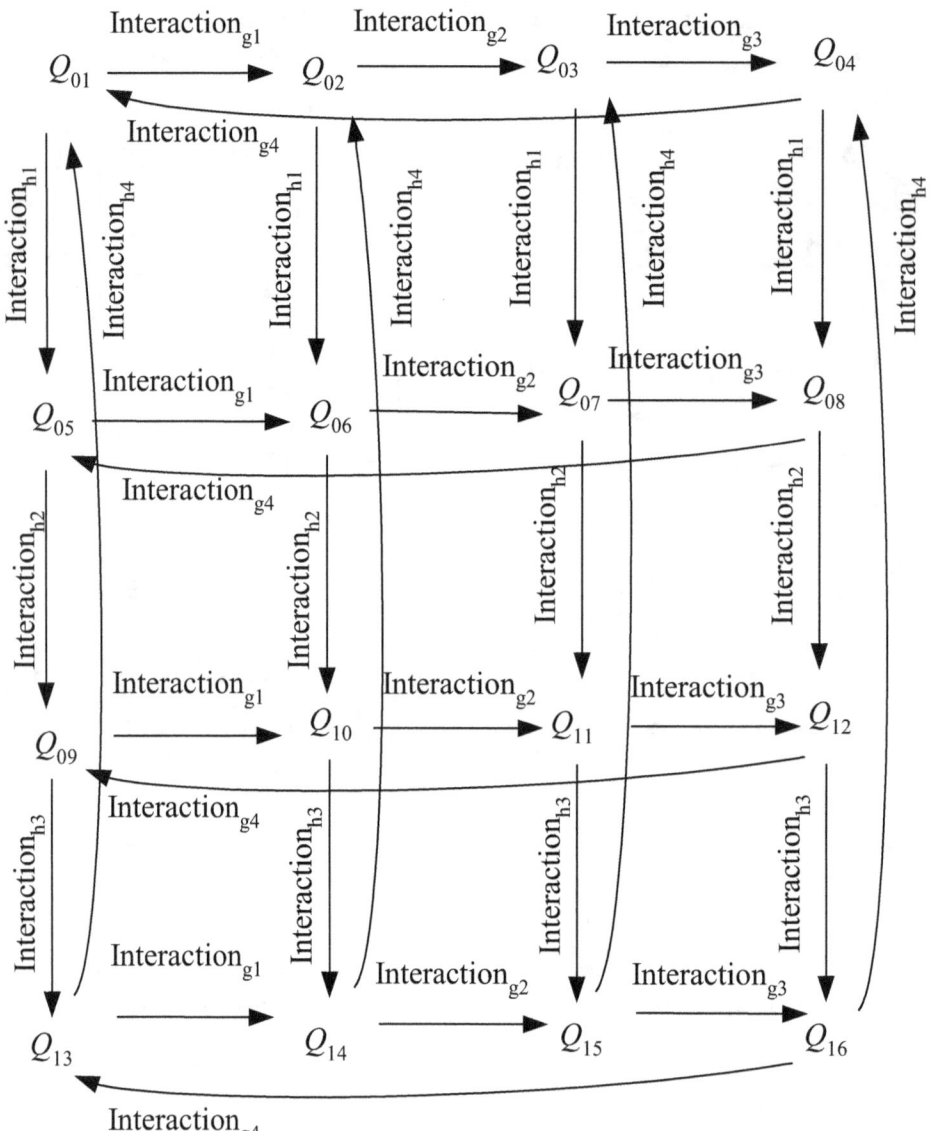

Figure 15-102   Transition Graph Demonstrates the Semantics of Process Q_{01}

We can easily verify that $S = \{(P_{01}, Q_{01}), (P_{02}, Q_{02}), (P_{03}, Q_{03}), (P_{04}, Q_{04}), (P_{05}, Q_{05}), (P_{06}, Q_{06}), (P_{07}, Q_{07}), (P_{08}, Q_{08}), (P_{09}, Q_{09}), (P_{10}, Q_{10}), (P_{11}, Q_{11}), (P_{12}, Q_{12}), (P_{13}, Q_{13}), (P_{14}, Q_{14}), (P_{15}, Q_{15}), (P_{16}, Q_{16})\}$ is a bisimulation.

Using the S bisimulation, we then are able to verify that P_{01} and Q_{01} are observation congruent because (1) $P_{01} \xrightarrow{\text{Interaction}_{g1}} P_{02}$, then we have Q_{02} that $Q_{01} \xRightarrow{\text{Interaction}_{g1}} Q_{02}$ and $P_{02} \overset{\sim}{\approx} Q_{02}$, and (2) $P_{01} \xrightarrow{\text{Interaction}_{h1}} P_{05}$, then we have Q_{05} that $Q_{01} \xRightarrow{\text{Interaction}_{h1}} Q_{05}$ and $P_{05} \overset{\sim}{\approx} Q_{05}$, and (3) $Q_{01} \xrightarrow{\text{Interaction}_{g1}} Q_{02}$, then we have P_{02} that $P_{01} \xRightarrow{\text{Interaction}_{g1}} P_{02}$ and $P_{02} \overset{\sim}{\approx} Q_{02}$, and (4) $Q_{01} \xrightarrow{\text{Interaction}_{h1}} Q_{05}$, then we have P_{05} that $P_{01} \xRightarrow{\text{Interaction}_{h1}} P_{05}$ and $P_{05} \overset{\sim}{\approx} Q_{05}$.

$P_{01} = Q_{01}$ means that "the design view of *Kurdi University*" and "the structural composition of the implementation view of *Kurdi University*" are observation congruent.

That is, we have demonstrated that the design view of *Kurdi University* is one level up structural composition (with observation congruence verification) of the implementation view of *Kurdi University*.

PART IV:
EVOLUTION&MOTIVATION
VIEW

Chapter 16: Higher-Order Systems

Higher-order systems interact with the environment through the exchange of not only matter, energy, data, information, or message but also systems. In this chapter, after introducing higher-order functions, second-order logic and higher-order systems, we then will give some examples of higher-order systems.

16-1 Higher-Order Functions

In mathematics, a higher-order function is a function that does at least one of the following: a) takes one or more functions as an input, b) outputs a function [Bare84, Hend80, Sang03]. All other functions are first-order functions.

A first-order function takes a combination of two (or more) sets of data to a single set of data, as shown in Figure 16-1. In the figure, input a, b and output c are data.

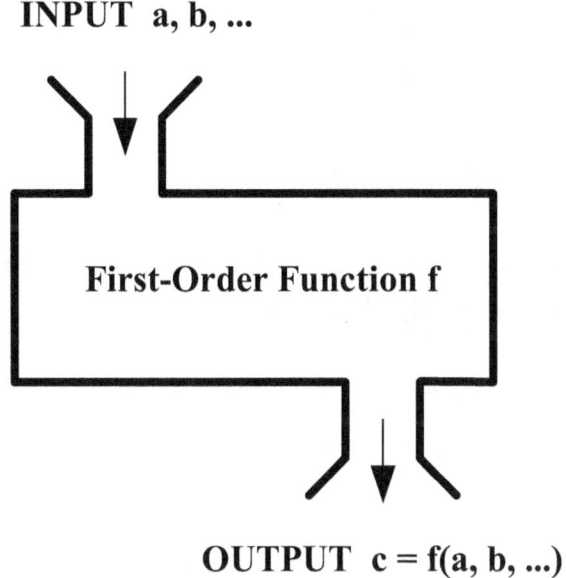

INPUT a, b, ...

First-Order Function f

OUTPUT c = f(a, b, ...)

Figure 16-1   First-Order Function

A higher-order function takes functions as input and return functions as output [Bare84, Hend80, Sang03], as shown in Figure 16-2. In the figure, input f_1, f_2 and output f_3 are functions.

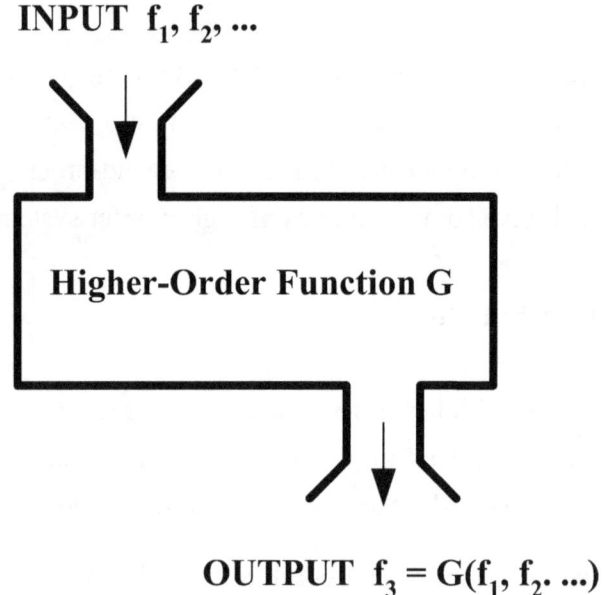

INPUT $f_1, f_2, \ldots$

Higher-Order Function G

OUTPUT $f_3 = G(f_1, f_2, \ldots)$

Figure 16-2   Higher-Order Function

For a higher-order function is computable, it must be monotonic and continuous [Bare84, Cohe63, Mann74, Scot67].

16-2 Second-Order Logic

In logic, second-order logic is an extension of first-order logic, which itself is an extension of propositional logic [Mann74, Shap00].

First-order logic quantifies only variables that range over elements of the domain of discourse, as shown in Figure 16-3.

$$\forall x \, \exists y \, (x \longrightarrow P\,(y))$$

Figure 16-3   First-Order Logic

In contrast, second-order logic, in addition to first-order logic, also quantifies over relations [Mann74, Shap00], as shown in Figure 16-4.

$$\forall Q \; \forall x \; \exists y \; (x \longrightarrow Q(y))$$

Figure 16-4   Second-Order Logic

For a second-order logic is computable, it must be monotonic and continuous [Bare84, Mann74, Scot67].

16-3 Higher-Order Systems

First-order systems interact with the environment through the exchange of matter, energy, data, information, or message, as shown in Figure 16-5.

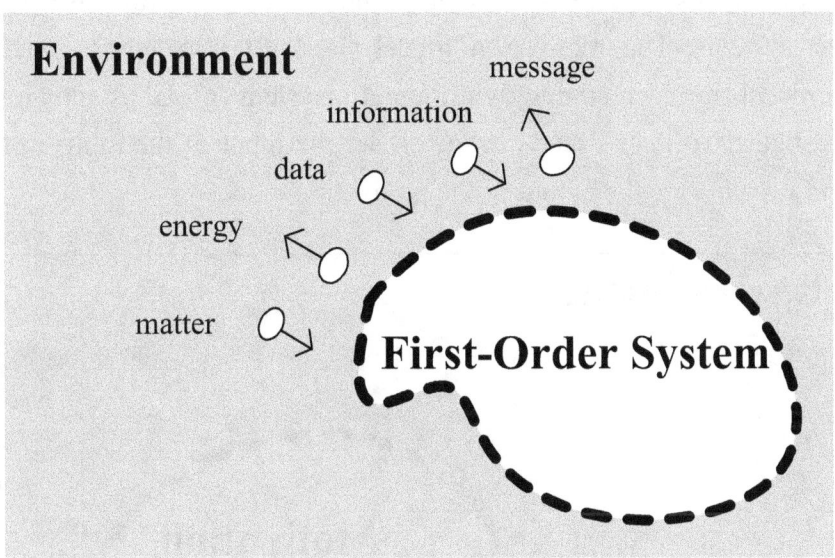

Figure 16-5   First-Order System

Higher-order systems interact with the environment through the exchange of not only matter, energy, data, information, or message but also systems, as shown in Figure 16-6.

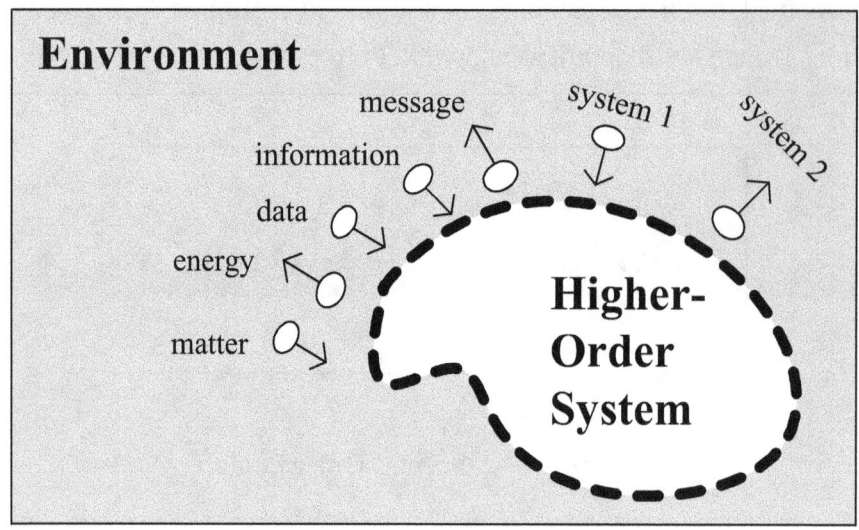

Figure 16-6   Higher-Order System

For a higher-order system is computable, it must be monotonic and continuous [Bare84, Mann74, Scot67].

16-4 Examples of Higher-Order Systems

Motivation model, creative thinking, and system dynamics are regarded as higher-order systems. The motivation model [Berk08] is a higher-order system. Motivation model, for each strategy will output a system (goal), as shown in Figure 16-7. In the figure, *strategy 1* and *strategy n* are the input of the motivation model; *system 1* and *system n* are the output of the motivation model

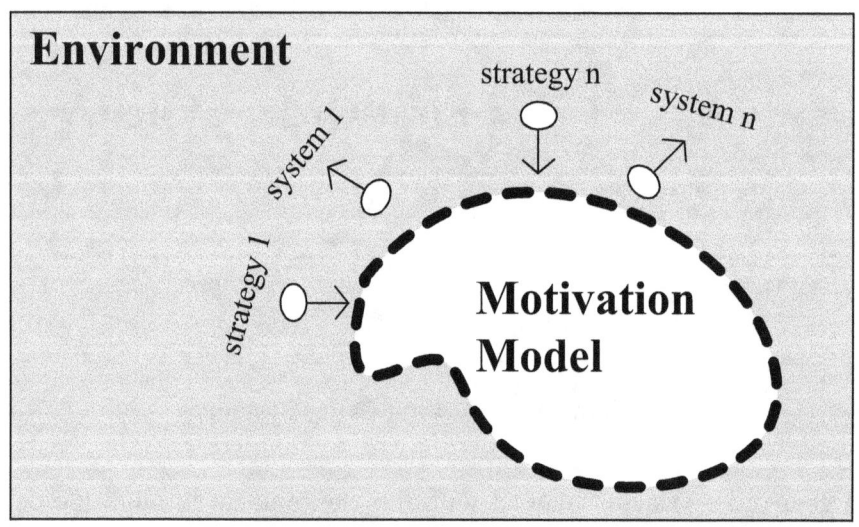

Figure 16-7   Motivation Model is a Higher-Order System

Creative thinking is a higher-order system, because it will be creating a large number of systems then chooses the best one, as shown in Figure 16-8. In the figure, *system 1*, *system* 2, and *system n* are the output of the creative thinking.

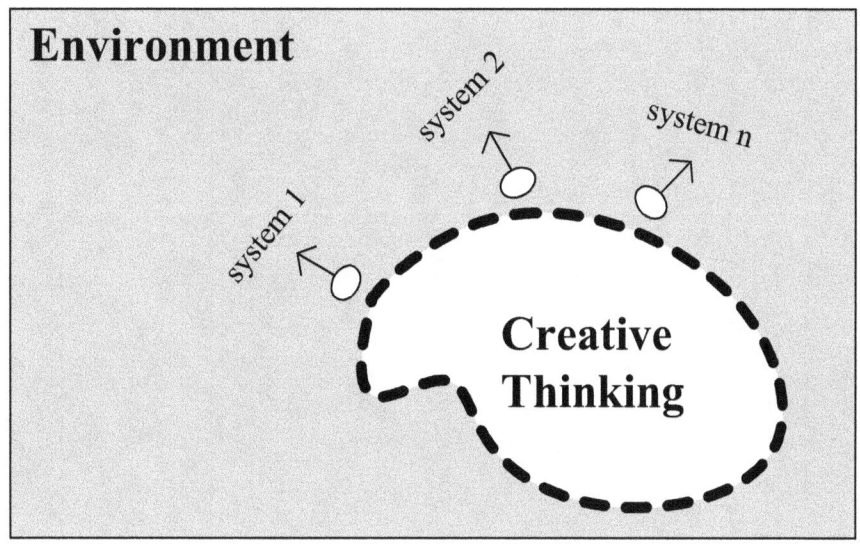

Figure 16-8   Creative Thinking is a Higher-Order System

System dynamics [Forr61, Ogat03, Palm09] is also a higher-order system, because it dynamically simulates the causal relationship among a large number of systems as shown in Figure 16-9. From these simulated systems, decision makers thus are able to strategically choose the most appropriate one.

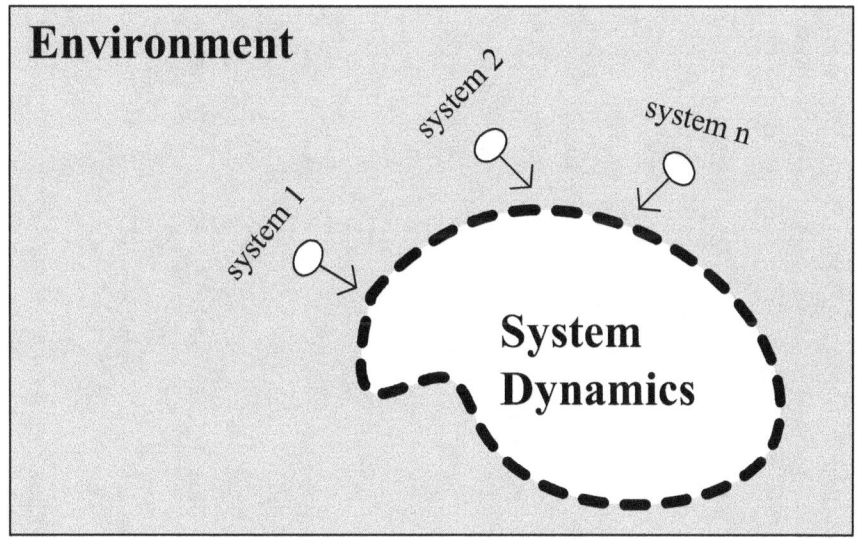

Figure 16-9   System Dynamics is a Higher-Order System

Chapter 17: Strategic Management

Strategic management is concerned primarily with responses to external issues such as in understanding the needs of customers and responding to competitive forces.

Strategic management provides overall direction to the enterprise. In short, it entails specifying the organizations goals, developing means designed to achieve these goals, and then allocating resources to implement the means.

This chapter first introduces what a strategy is. Then it discusses the motivation model is a higher-order system. Last, this chapter will work on the strategic means.

17-1 Strategy

Strategy is included in a high level plan to achieve one or more goals under conditions of uncertainty [Free13, Mcke12]. Strategy is important because the resources available to achieve these goals are usually limited.

In the business motivation model [Berk08], as shown in Figure 17-1, strategy is the human attempt to get to "desirable ends with available means".

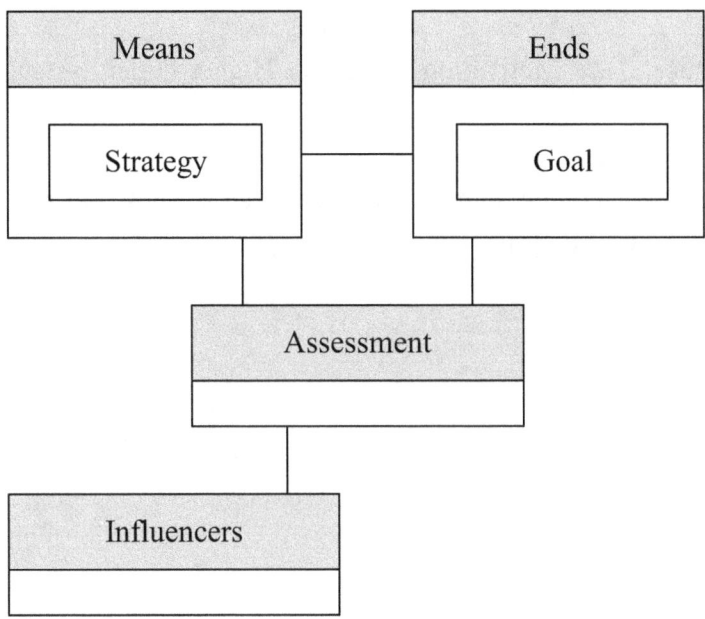

Figure 17-1   Business Motivation Model

In the above figure, available means are related with the strategy; desirable ends are related with the goal.

17-2 Motivation Model is a Higher-Order System

As discussed in the previous chapter, motivation model is a higher-order system. Motivation model will output a system (goal) for each strategy as shown in Figure 17-2.

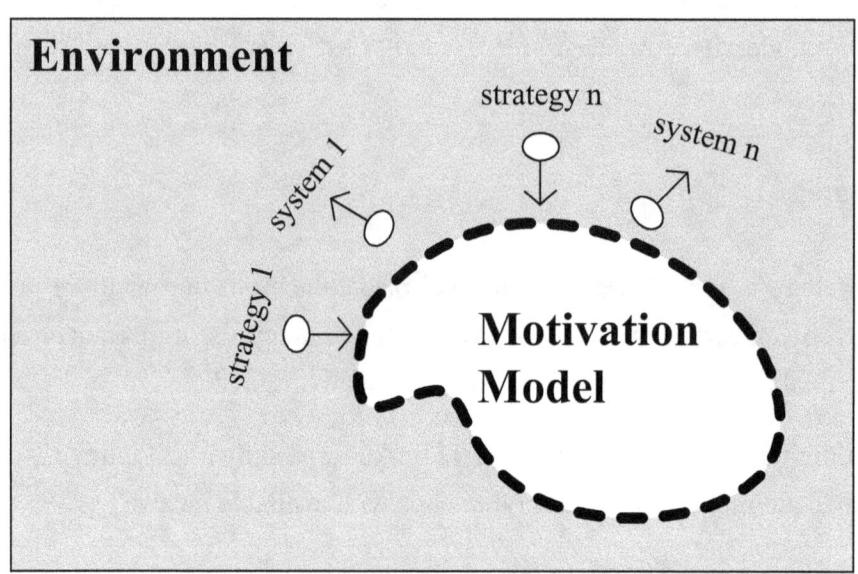

Figure 17-2   Motivation Model is a Higher-Order System

In general, each strategy is mapped to a system as shown in Figure 17-3. That is, the motivation model will output a system for each strategy.

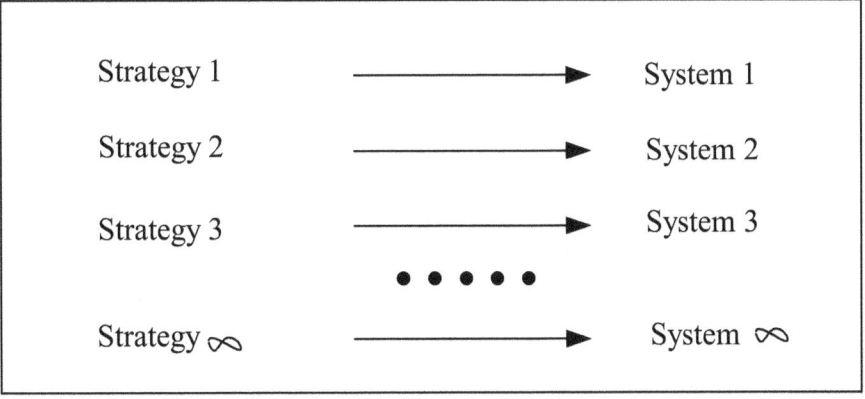

Figure 17-3   Each Strategy is Mapped to a System

17-3 Strategic Means

In the business motivation model, strategy is the human attempt to get to "desirable ends with available means". Strategic means analyzes the major initiatives taken by a company's top management on behalf of business owners, involving resources and performance in internal and external environments.

Strategic means include: (a) goal drivers, (b) goal assumptions, (c) goal constraints, and (d) SWOT (strengths, weaknesses, opportunities, threat) analysis, etc. We use these strategic means to achieve the desirable ends.

Goal drivers are up from the policy considerations, the goal driver is kind of why we want to have those desirable ends. Goal assumptions are taking into account of those assumptions that have a positive impact on these desirable ends. Goal constraints are up from the policy considerations, the goal constraints are related to those restrictions which have a negative impact on those desirable ends. SWOT analysis is to analyze the internal strengths, weaknesses, opportunities, and threats, and so for executing this strategy.

Chapter 18: Evolution of a System

A system, not matter it is physical or virtual, will always change from time to time. A system evolves when it changes. This chapter introduces the evolution of a system.

18-1 Change from the Internal Force

The change cause may come from the internal or external forces of a system. Cell division, as shown in Figure 18-1, is an example of change from the internal force.

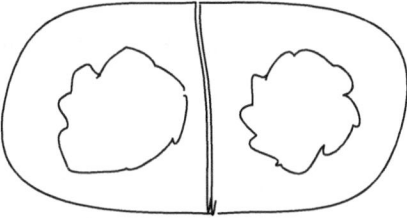

Figure 18-1   Cell Division

As a second example, a company transforms itself to a desired state, as shown in Figure 18-2, is the other instance of change from the internal force.

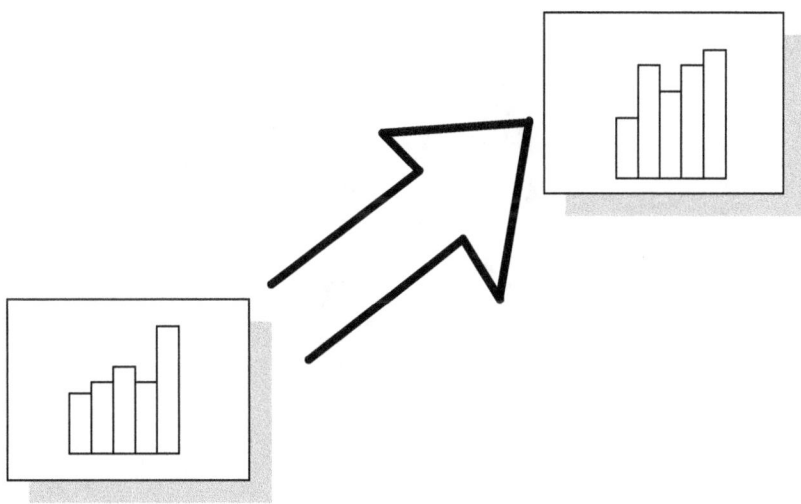

Figure 18-2   A Company Transforms Itself

As a third example, a killer transforms himself into a good man, as shown in Figure 18-3, is another instance of change from the internal force.

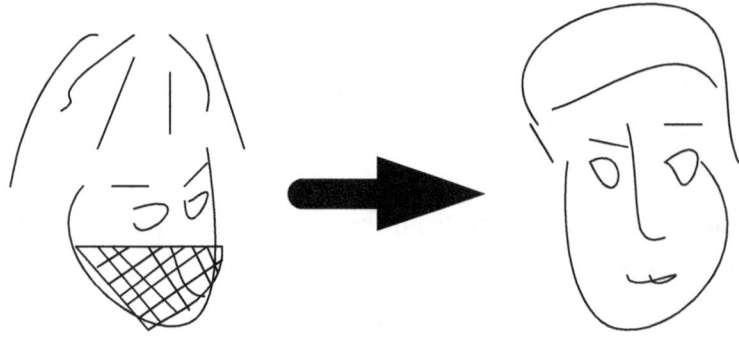

Figure 18-3　A Killer Transforms Himself

18-2 Change from the External Force

The change cause may come from the internal or external forces of the system. Repair workers doing the tire rotation, as shown in Figure 18-4, is an example of change from the external force.

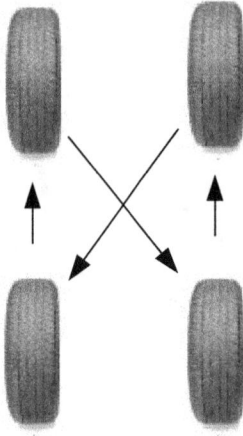

Figure 18-4　Tire Rotation

As a second example, workers doing road maintenance as shown in Figure 18-5, is the other instance of change from the external force.

Figure 18-5   Road Maintenance

As a third example, remodeling a house as shown in Figure 18-6, is another instance of change from the external force.

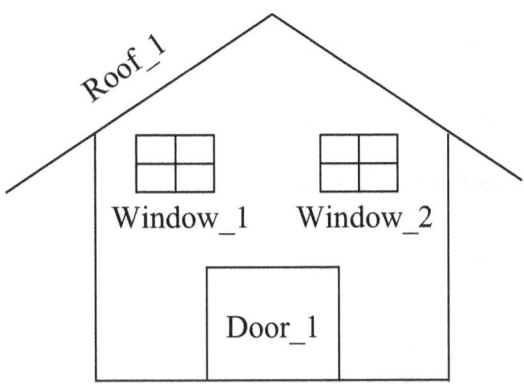

Figure 18-6   Remodeling a House

18-3 Result of Systems Evolution

A system evolves when it changes. Evolution of a system is shown in Figure 18-7.

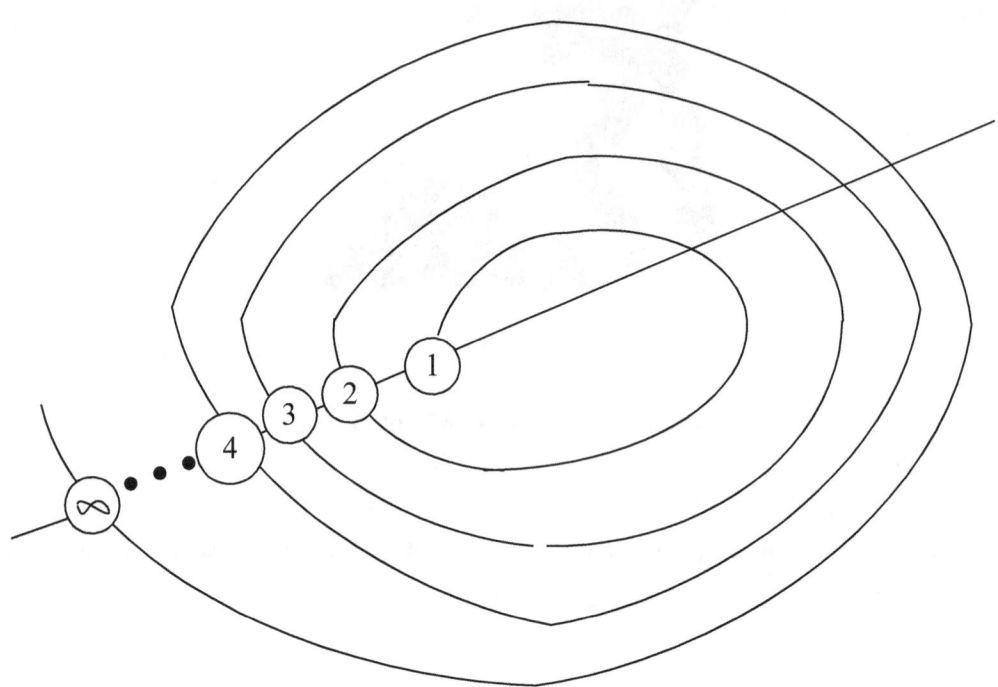

① : Systems Definition Version 1

② : Systems Definition Version 2

③ : Systems Definition Version 3

④ : Systems Definition Version 4

• • •

∞ : Systems Definition Version ∞

Figure 18-7   Evolution of a System

Each time when a system changes or evolves, we shall get a new version of its systems definition as the result of systems evolution. In the above figure, *version 1* stands for the original systems definition of a system and evolves into the *version 2*, *version 3*,…, and *version ∞* gradually.

According the motivation model, one strategy is mapped to one version for any systems definition, as shown in Figure 18-8.

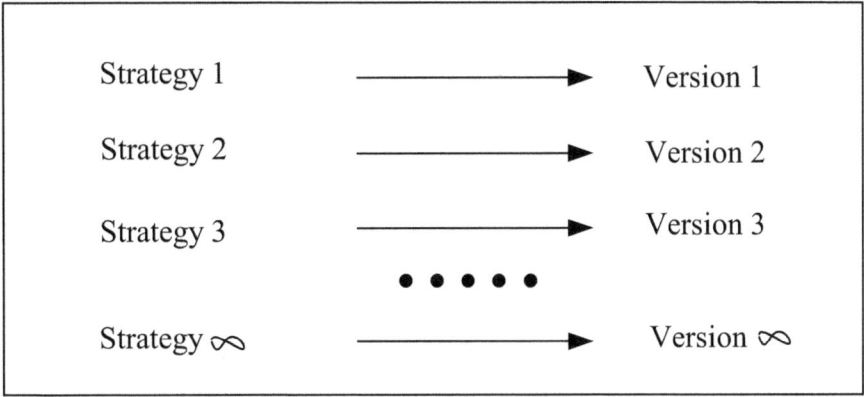

Figure 18-8   One Strategy is Mapped to One Version

Evolution of a system is represented as the SBC evolution&motivation view shown in Figure 18-9. In the figure, we see that, for any systems definition, one strategy is mapped to one version.

Dimension 1:
Evolution&Motivation View

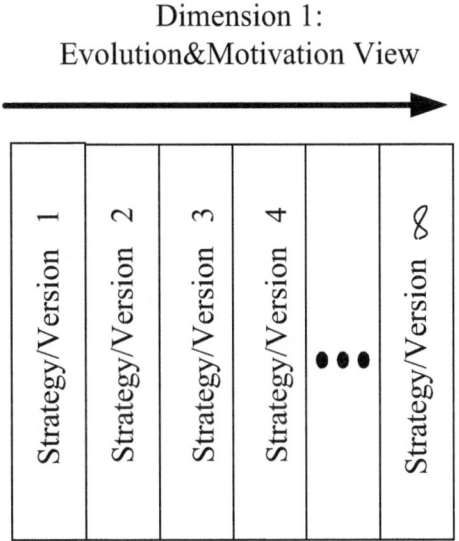

Figure 18-9   SBC Evolution&Motivation View

Chapter 19: Multi-Queue SBC Process Algebra Language Constructs Regarding the Evolution&Motivation View

In the chapter, we illustrate in detail those multi-queue SBC process algebra language constructs which make up the evolution&motivation view.

19-1 Backus-Naur Form for the Evolution&Motivation View

The set of multi-queue SBC (i.e. Structure-Behavior Coalescence) process for the evolution&motivation view is defined by the following BNF grammar, as shown in Figure 19-1.

    (1) <Multi-Queue_SBC_Process_of_Evolution&Motivation_View> ::=
        <FixIFD>

    (2) <FixIFD> ::=
        fix("<Process_Variable>"="<IFD> " $\bullet$ " <Process_Variable>")"

    (3) <IFD>   ::=
        <Type_1_Interaction>

    (4) <Type_1_Interaction> ::= <Actor> <Operation_Call_Or_Return>
            <Operation_Call_Or_Return_Formula> <Component>

Figure 19-1   Backus-Naur Form for the Evolution&Motivation View

19-2 A Recursive Interaction Flow Diagram Defines the Multi-Queue SBC Process of the Evolution&Motivation View

Rule 1 describes that a recursive interaction flow diagram (i.e. FixIFD) defines the multi-queue SBC process of the evolution&motivation view, as shown in Figure 19-2.

Rule 1
<Multi-Queue_SBC_Process_of_Evolution&Motivation_View> ::=   <FixIFD>

Figure 19-2   Rule 1

19-3 Recursion of an Interaction Flow Diagram Defines the Recursive Interaction Flow Diagram

Rule 2 describes that we use the recursion (i.e. **fix**) of an interaction flow diagram (i.e. IFD) to define a recursive interaction flow diagram (i.e. FixIFD), as shown in Figure 19-3.

Rule 2
<FixIFD> ::=   **fix**(”<Process_Variable>“=”<IFD> “● ”<Process_Variable>“)”

Figure 19-3   Rule 2

19-4 A Type_1 Interaction Defines an Interaction Flow Diagram

Rule 3 describes that a type_1 interaction defines an interaction flow diagram (i.e. IFD), as shown in Figure 19-4.

Rule 3
<IFD>   ::=   <Type_1_Interaction>

Figure 19-4   Rule 3

19-5 An Actor Interacting with a Component Defines the Type_1 Interaction

Rule 4 describes that an actor interacting with a component defines the type_1 interaction, as shown in Figure 19-5.

Rule 4
<Type_1_Interaction> ::= <Actor> <Operation_Call_Or_Return>   <Operation_Call_Or_Return_Formula> <Component>

Figure 19-5   Rule 4

Chapter 20: Multi-Queue SBC Process Algebra Transitional Semantics Regarding the Evolution&Motivation View

In the chapter, we illustrate in detail those multi-queue SBC process algebra transitional semantics which regards the evolution&motivation view.

20-1 Transitional Semantics for the Evolution&Motivation View

As shown in Figure 20-1, we assume an infinite set G of type_1 interactions, and use g_1, g_2...to range over G. Further, we let X be the set of process variables, and use X_1, X_2...to range over X. We let Φ be the set of process Constants, and use A_1, A_2...to range over Φ. We let Π be the set of processes, and use P_1, Q_1...to range over Π. We let Ψ be the set of process expressions, and use E_1, E_2...to range over Ψ.

Entity set	Entity name	Type of entity
G	$g_1, g_2...$	type_1 interactions
X	$X_1, X_2...$	process variables
Φ	$A_1, A_2...$	process Constants
Π	$P_1, Q_1...$	processes
Ψ	$E_1, E_2...$	process expressions

Figure 20-1   Entities

In giving meaning to the multi-queue SBC process algebra for the evolution&motivation view, we shall use the following labelled transition system (LTS) [Miln89, Miln99]

$$(\Psi, G, \rightarrow)$$

which consists of a set Ψ of process expressions, a set G of "type_1 interactions", and a transition relation $\rightarrow \subseteq \Psi \times G \times \Psi$ where $(E_i, g, E_j) \in \rightarrow$ is denoted by $E_i \xrightarrow{g} E_j$.

The semantics for Ψ consists in the transition rules of each transition relation $\rightarrow$ over $\Psi \times G \times \Psi$. These transition rules will follow the structure of expressions.

As shown in Figure 20-2, we give the complete set of transition rules; the names Prefix, Recursion, and Constant indicate that the rules are associated respectively with Prefix and Recursion and with Constants.

<table>
<tr><td>Prefix</td><td>$$\frac{}{g \bullet E \xrightarrow{g} E}$$</td></tr>
<tr><td>Recursion</td><td>$$\frac{\mathbf{fix}(X = z\{\mathbf{fix}(X=z)/X\}) \xrightarrow{g} E'}{\mathbf{fix}(X=z) \xrightarrow{g} E'}$$</td></tr>
<tr><td>Constant</td><td>$$\frac{P \xrightarrow{g} P'}{A \xrightarrow{g} P'} \quad (A \stackrel{\text{def}}{=\!=} P)$$</td></tr>
</table>

Figure 20-2   Transition Rules for the Evolution&Motivation View

20-2 Rule of Prefix

The rule for Prefix, shown in Figure 20-3, can be read as follows: Under any circumstances, we always infer $a \bullet E \xrightarrow{g} E$. That is, an expression, with an interaction prefixed to it, will use this interaction to accomplish the transition.

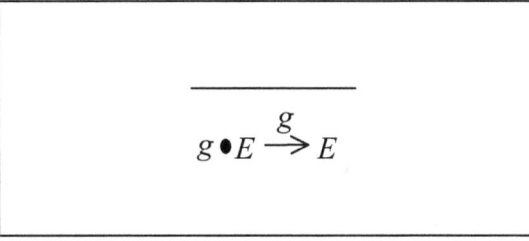

Figure 20-3   Rule of Prefix

20-3 Rule of Recursion

The rule for Recursion, shown in Figure 20-4, can be read as follows: This says that any interaction which may be inferred for the **fix** expression 'unwound' once (by substituting itself for its bound variable) may be inferred for the **fix** expression itself.

$$\mathbf{fix}(X{=}z\{\mathbf{fix}(X{=}z)/X\}) \xrightarrow{g} E'$$
$$\mathbf{fix}(X{=}z) \xrightarrow{g} E'$$

Figure 20-4   Rule of Recursion

20-4 Rule of Constants

The rule for Constants, shown in Figure 20-5, can be read as follows: the rule of Constants asserts that each Constant has the same transitions as its defining expression.

$$\frac{P \xrightarrow{g} P'}{A \xrightarrow{g} P'} \quad (A \overset{\text{def}}{=} P)$$

Figure 20-5   Rule of Constants

Chapter 21: Kurdi University as an Example

In the SBC evolution&motivation view of *Kurdi University*, we see that, for its systems definition, one strategy is mapped to one version as shown in Figure 21-1.

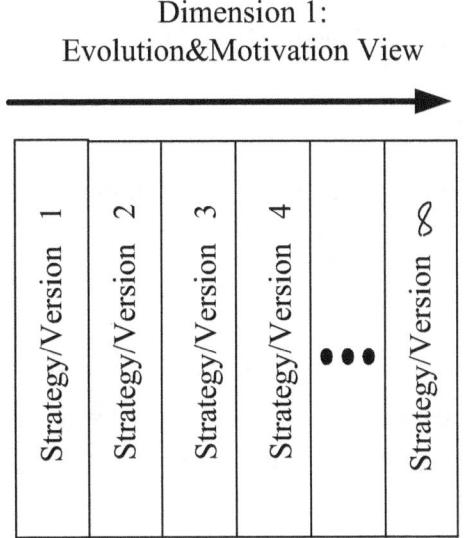

Figure 21-1   SBC Evolution&Motivation View

21-1 Overall Behavior of the Evolution&Motivation View of Kurdi University

The overall behavior of the evolution&motivation view of *Kurdi University* includes one behavior: *Motivation_Modeling* as shown in Figure 21-2. This behavior is described by an individual IFD.

228

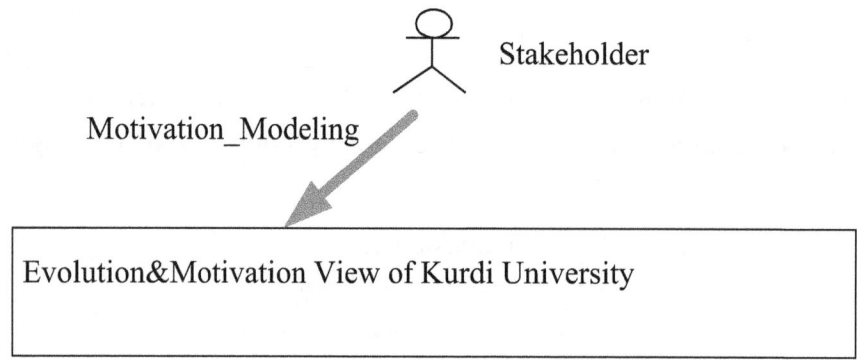

Figure 21-2   Overall Behavior of the Evolution&Motivation View
of Kurdi University

An IFD of the *Motivation_Modeling* behavior is shown in Figure 21-3. First, actor *Stakeholder* interacts with the *Motivation_Model* component through the *Motivation* operation call interaction, carrying the *Strategy* input parameter and *Version* output parameter.

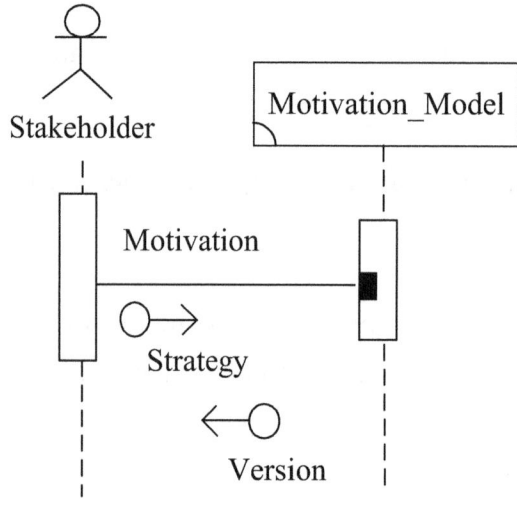

Figure 21-3   IFD of the *Motivation_Modeling* Behavior

21-2 Backus-Naur Form for the Evolution&Motivation View of Kurdi University

We draw the multi-queue SBC process algebra Backus-Naur Form tree of the evolution&motivation view of *Kurdi University* as shown in Figure 21-4.

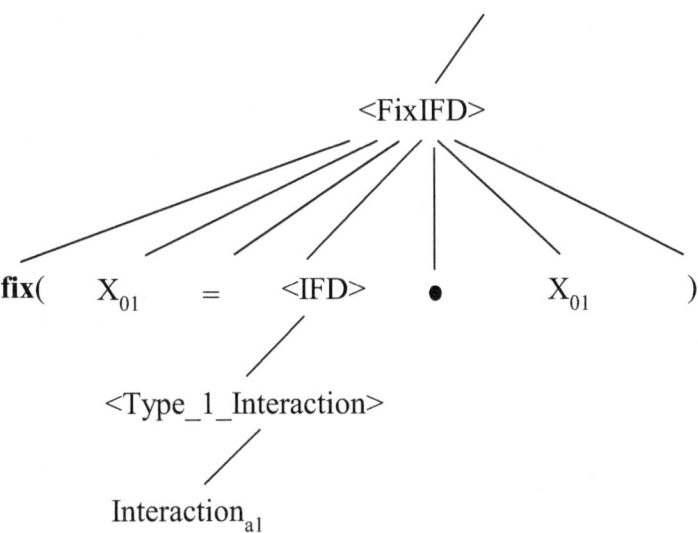

<Kurdi University's Evolution&Motivation View>

<FixIFD>

fix(X_{01} = <IFD> • X_{01})

<Type_1_Interaction>

$Interaction_{a1}$

Figure 21-4 BNF Tree of the Evolution&Motivation View
of Kurdi University

21-3 Interactions of the Evolution&Motivation View of Kurdi University

$Interaction_{a1}$ stands for the 1st interaction of the ath interaction flow diagram of the evolution&motivation view of *Kurdi University*, as shown in Figure 21-5. $Interaction_{a1}$ is a type_1 interaction which describes the *Stakeholder* actor interacts with the *Motivation_Model* component.

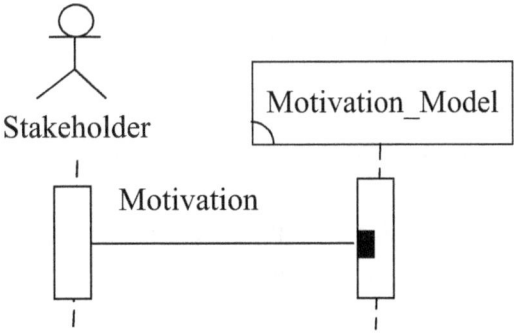

Stakeholder

Motivation_Model

Motivation

Figure 21-5 $Interaction_{a1}$

21-4 Interaction Flow Diagrams of the Evolution&Motivation View of Kurdi University

FixIFD$_a$ describes the recursion of the ath interaction flow diagram of the evolution&motivation view of *Kurdi University*. FixIFD$_a$ is syntactically represented as "$\textbf{fix}(X_{01}=\text{Interaction}_{a1} \bullet X_{01})$", as shown in Figure 21-6.

$$\text{FixIFD}_a \overset{\text{def}}{=\!=} $$

$$\textbf{fix}(X_{01} = \text{Interaction}_{a1} \bullet X_{01})$$

Figure 21-6　FixIFD$_a$

21-5 Multi-Queue SBC Process of the Evolution&Motivation View of Kurdi University

The multi-queue SBC process of the evolution&motivation view of *Kurdi University* is syntactically represented as FixIFD$_a$ which equals to "$\textbf{fix}(X_{01}=\text{Interaction}_{a1} \bullet X_{01})$", as shown in Figure 21-7.

$$\text{Multi-Queue SBC Process of the Evolution\&Motivation View of Kurdi University} \overset{\text{def}}{=\!=}$$

$$\textbf{fix}(X_{01} = \text{Interaction}_{a1} \bullet X_{01})$$

Figure 21-7　Multi-Queue SBC Process of the Evolution&Motivation View of Kurdi University

APPENDIX A: SBC VIEW MODEL (SBC-VM)

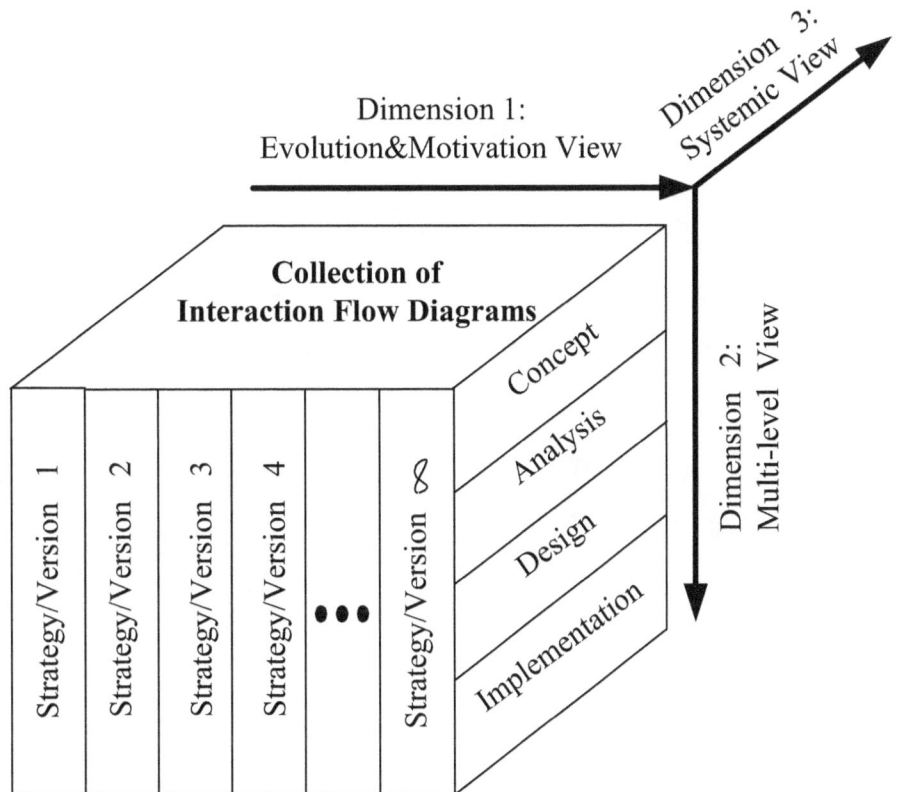

APPENDIX B: SBC ARCHITECTURE DEVELOPMENT METHOD (SBC-ADM)

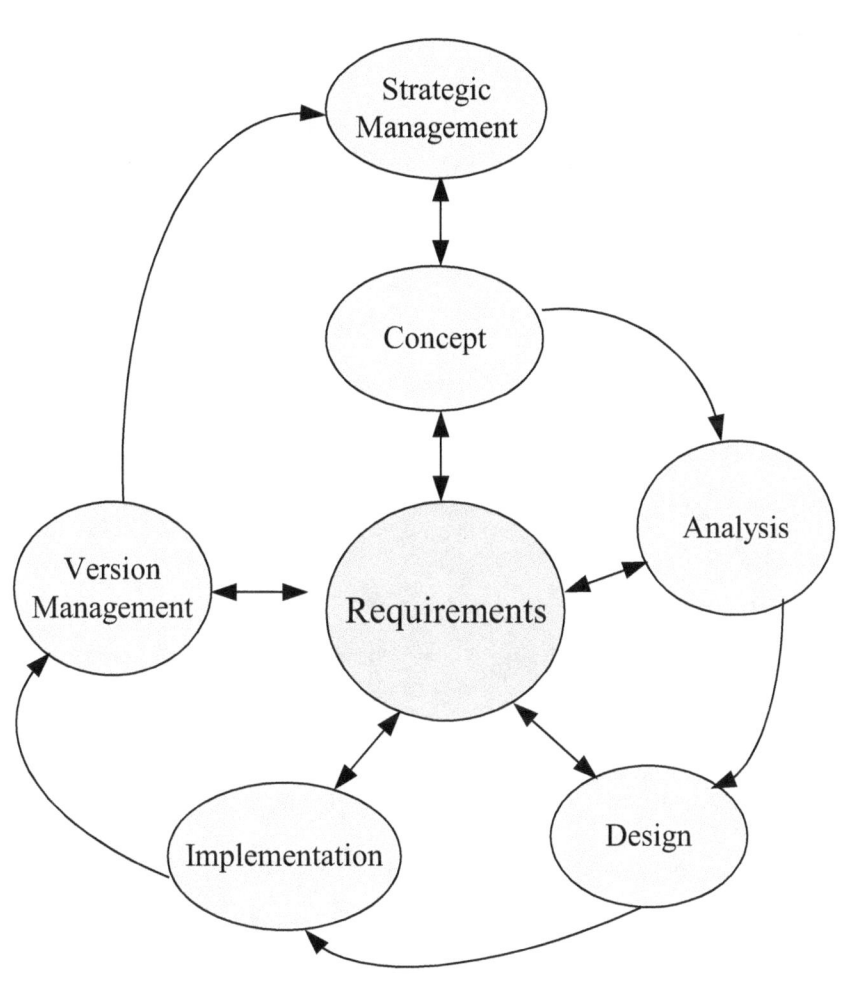

APPENDIX C: MULTI-QUEUE SBC PROCESS ALGEBRA (M-SBC-PA)

(1) Systemic View

(1) <Multi-Queue_SBC_Process_of_Systemic_View> ::=
    <Parallel_of_FixIFD>

(2) <Parallel_of_FixIFD> ::=   *STOP*
  | <FixIFD> " ∥ " <Parallel_of_FixIFD>

(3) <FixIFD> ::= "**fix**("<Process_Variable>"="
        <IFD><Process_Variable>")"

(4) <IFD> ::=
  <Type_1_Expression> <Zero_Or_More_Expressions>

(5) <Zero_Or_More_Expressions> ::= " ● "
  | " ● " <Type_1_Or_2_Expression> <Zero_Or_More_Expressions>

(6) <Type_1_Or_2_Expression> ::= <Type_1_Expression>
  | <Type_2_Expression>

(7) <Type_1_Expression> ::= <Type_1_Interaction>
  | <Condition> <Type_1_Interaction>
    {"+" <Condition> <Type_1_Interaction>}

(8) <Type_2_Expression> ::= <Type_2_Interaction>
  | <Condition> <Type_2_Interaction>
    {"+" <Condition> <Type_2_Interaction>}

(9) <Type_1_Interaction> ::= <Actor> <Operation_Call_Or_Return>
    <Operation_Call_Or_Return_Formula> <Component>

(10) <Type_2_Interaction> ::= <Component> <Operation_Call_Or_Return>
    <Operation_Call_Or_Return_Formula> <Component>

Prefix
$$\frac{\qquad\qquad}{a \bullet E \xrightarrow{a} E}$$

Parallel$_1$
$$\frac{E \xrightarrow{a} E'}{E \parallel F \xrightarrow{a} E' \parallel F}$$

Parallel$_2$
$$\frac{F \xrightarrow{a} F'}{E \parallel F \xrightarrow{a} E \parallel F'}$$

Recursion
$$\frac{\mathbf{fix}(X=z\{\mathbf{fix}(X=z)/X\}) \xrightarrow{a} E'}{\mathbf{fix}(X=z) \xrightarrow{a} E'}$$

Constant
$$\frac{P \xrightarrow{a} P'}{A \xrightarrow{a} P'} \quad (A \stackrel{\text{def}}{=} P)$$

(2) Multi-Level View

(1) <Structural Composition of Multi-Queue_SBC_Process_of_Each_Level's_Systemic_View> ::=
      <Multi-Queue_SBC_Process_of_Each_Level's_Systemic_View>[*f*]

(2) <Multi-Queue_SBC_Process_of_Each_Level's_Systemic_View> ::=
      <Parallel_of_FixIFD>

(3) <Parallel_of_FixIFD> ::=   *STOP*
  | <FixIFD> " ║ " <Parallel_of_FixIFD>

(4) <FixIFD> ::= **fix**("<Process_Variable>"="
                     <IFD><Process_Variable>")"

(5) <IFD> ::=
    <Type_1_Expression> <Zero_Or_More_Expressions>

(6) <Zero_Or_More_Expressions> ::= " ● "
  | " ● " <Type_1_Or_2_Expression> <Zero_Or_More_Expressions>

(7) <Type_1_Or_2_Expression> ::= <Type_1_Expression>
  | <Type_2_Expression>

(8) <Type_1_Expression> ::= <Type_1_Interaction>
  | <Condition> <Type_1_Interaction>
    {"+" <Condition> <Type_1_Interaction>}

(9) <Type_2_Expression> ::= <Type_2_Interaction>
  | <Condition> <Type_2_Interaction>
    {"+" <Condition> <Type_2_Interaction>}

(10) <Type_1_Interaction> ::= <Actor> <Operation_Call_Or_Return>
          <Operation_Call_Or_Return_Formula> <Component>

(11) <Type_2_Interaction> ::= <Component> <Operation_Call_Or_Return>
          <Operation_Call_Or_Return_Formula> <Component>

Prefix

$$\frac{}{\alpha \bullet E \xrightarrow{\alpha} E}$$

Parallel₁

$$\frac{E \xrightarrow{\alpha} E'}{E \parallel F \xrightarrow{\alpha} E' \parallel F}$$

Parallel₂

$$\frac{F \xrightarrow{\alpha} F'}{E \parallel F \xrightarrow{\alpha} E \parallel F'}$$

Recursion

$$\frac{\mathbf{fix}(X=z\{\mathbf{fix}(X=z)/X\}) \xrightarrow{\alpha} E'}{\mathbf{fix}(X=z) \xrightarrow{\alpha} E'}$$

Rename

$$\frac{E \xrightarrow{\alpha} E'}{E[f] \xrightarrow{\alpha[f]} E'[f]}$$

Constant

$$\frac{P \xrightarrow{\alpha} P'}{A \xrightarrow{\alpha} P'} \quad (A \stackrel{\mathrm{def}}{=} P)$$

(3) Evolution&Motivation View

(1) <Multi-Queue_SBC_Process_of_Evolution&Motivation_View> ::=
    <FixIFD>

(2) <FixIFD> ::=
    "**fix***(*"<Process_Variable>"="<IFD> " ● " <Process_Variable>"*)*"

(3) <IFD>   ::=
    <Type_1_Interaction>

(4) <Type_1_Interaction> ::= <Actor> <Operation_Call_Or_Return>
    <Operation_Call_Or_Return_Formula> <Component>

$$\text{Prefix} \qquad \dfrac{\rule{3cm}{0.4pt}}{g \bullet E \xrightarrow{g} E}$$

$$\text{Recursion} \qquad \dfrac{\mathbf{fix}(X=z\{\mathbf{fix}(X=z)/X\}) \xrightarrow{g} E\,'}{\mathbf{fix}(X=z) \xrightarrow{g} E\,'}$$

$$\text{Constant} \qquad \dfrac{P \xrightarrow{g} P\,'}{A \xrightarrow{g} P\,'} \quad (A \overset{\text{def}}{=\!=} P)$$

BIBLIOGRAPHY

[Acko68] Ackoff, R., "Toward a System of Systems Concepts," *Modern Systems Research for the Behavioral Scientist: A Sourcebook*, Aldine Publishing Company, 1968.

[Bare84] Barendregt, H. P., *The Lambda Calculus: Its Syntax and Semantics*, Elsevier Science Publishers, 1984.

[Beam90] Beam, W. R., *Systems Engineering: Architecture and Design*, McGraw-Hill, 1990.

[Bere09] Berenbach, B. et al., *Software & Systems Requirements Engineering: In Practice*, 1st Edition, McGraw-Hill Osborne Media, 2009.

[Berg87] Bergstra, J. A. et al., "ACPτ: A Universal Axiom System for Process Specification," *CWI Quarterly* 15, 1987, pp. 3-23.

[Berk08] Berkem, B., "From the Business Motivation Model to Service Oriented Architecture," *Journal of Object Technology*, Vol.7, No.8, 2008.

[Bert69] Von Bertalanffy, L., *General System Theory: Foundations, Development, Applications*, Revised Edition, George Braziller Inc., 1969.

[Bert81] Von Bertalanffy, L. et al., *Systems View of Man: Collected Essays*, Westview Pr, 1981.

[Burd10] Burd, S. D., *Systems Architecture*, 6th Edition, Cengage Learning, 2010.

[Chao14a] Chao, W. S., *Systems Thingking 2.0: Architectural Thinking Using the SBC Architecture Description Language*, CreateSpace Independent Publishing Platform, 2014.

[Chao14b] Chao, W. S., *General Systems Theory 2.0: General Architectural Theory*

Using the SBC Architecture, CreateSpace Independent Publishing Platform, 2014.

[Chao14c] Chao, W. S., *Systems Modeling and Architecting: Structure-Behavior Coalescence for Systems Architecture*, CreateSpace Independent Publishing Platform, 2014.

[Chao15a] Chao, W. S., *Variants of Interaction Flow Diagrams: The Structure-Behavior Coalescence Approach*, CreateSpace Independent Publishing Platform, 2015.

[Chao15b] Chao, W. S., *A Process Algebra For Systems Architecture: The Structure-Behavior Coalescence Approach*, CreateSpace Independent Publishing Platform, 2015.

[Chao15c] Chao, W. S., *An Observation Congruence Model For Systems Architecture: The Structure-Behavior Coalescence Approach*, CreateSpace Independent Publishing Platform, 2015.

[Chao15d] Chao, W. S., *Variants of SBC Process Algebra: The Structure-Behavior Coalescence Approach*, CreateSpace Independent Publishing Platform, 2015.

[Chao15e] Chao, W. S., *Single-Queue SBC Process Algebra For Systems Architecture: The Structure-Behavior Coalescence Approach*, CreateSpace Independent Publishing Platform, 2015.

[Chao15f] Chao, W. S., *Multi-Queue SBC Process Algebra For Systems Architecture: The Structure-Behavior Coalescence Approach*, CreateSpace Independent Publishing Platform, 2015.

[Chao15g] Chao, W. S., *Single-Queue SBC Observation Congruence Model For Systems Architecture: The Structure-Behavior Coalescence Approach*, CreateSpace Independent Publishing Platform, 2015.

[Chao15h] Chao, W. S., *Multi-Queue SBC Observation Congruence Model For Systems Architecture: The Structure-Behavior Coalescence Approach*, CreateSpace Independent Publishing Platform, 2015.

[Chec99] Checkland, P., *Systems Thinking, Systems Practice: Includes a 30-Year Retrospective*, 1st Edition, Wiley, 1999.

[Cohe63] Cohen, P. J., "The Independence of the Continuum Hypothesis," *Proceedings of the National Academy of Sciences of the United States of America*, 50 (6), 1963, pp. 1143–1148.

[Dam06] Dam, S., *DoD Architecture Framework: A Guide to Applying System Engineering to Develop Integrated Executable Architectures*, BookSurge Publishing, 2006.

[Date03] Date, C. J., *An Introduction to Database Systems*, 8th Edition, Addison Wesley, 2003.

[Denn08] Dennis, A. et al., *Systems Analysis and Design*, 4th Edition, Wiley, 2008.

[Dori95] Dori, D., "Object-Process Analysis: Maintaining the Balance between System Structure and Behavior," *Journal of Logic and Computation* 5(2), pp.227-249, 1995.

[Dori02] Dori, D., *Object-Process Methodology: A Holistic Systems Paradigm*, Springer Verlag, New York, 2002.

[Dori16] Dori, D., *Model-Based Systems Engineering with OPM and SysML*, Springer Verlag, New York, 2016.

[Elma10] Elmasri, R., *Fundamentals of Database Systems*, 6th Edition, Addison Wesley, 2010.

[Forr61] Forrester, J. W., *Industrial Dynamics*, Pegasus Communications, 1961.

[Free13] Freeman, L., *Strategy: A History*, Oxford University Press, 2013.

[Frie11] Friedenthal, S., et al., *A Practical Guide to SysML, Second Edition: The*

Systems Modeling Language, 2nd Edition, Morgan Kaufmann, 2011.

[Gall03] Gall, J., *The Systems Bible: The Beginner's Guide to Systems Large and Small*, General Systemantics Pr/Liberty, 2003.

[Ghar11] Gharajedaghi, J., *Systems Thinking: Managing Chaos and Complexity: A Platform for Designing Business Architecture*, Morgan Kaufmann, 2011.

[Grad06] Grady, J. O., *System Requirements Analysis*, 1st Edition, Academic Press, 2006.

[Hend80] Henderson, P., *Functional Programming: Application and Implementation*, Prentice-Hall, 1980.

[Hoar85] Hoare, C. A. R., *Communicating Sequential Processes*, Prentice-Hall, 1985.

[Hoff10] Hoffer, J. A., et al., *Modern Systems Analysis and Design*, 6th Edition, Prentice Hall, 2010.

[Jorg12] Jorgensen, S. E., *Introduction to Systems Ecology (Applied Ecology and Environmental Management)*, CRC Press, 2012.

[Kapo94] Kaposi, A., et al., *Systems, Models and Measure*, Springer-Verlag London Limited, 1994.

[Kass07] Kasser, J. E., *A Framework for Understanding Systems Engineering*, BookSurge Publishing, 2007.

[Kill09] Killoran, D. M., *LSAT Logical Reasoning Bible: A Comprehensive System for Attacking the Logical Reasoning Section of the LSAT*, PowerScore Publishing, 2009.

[Klip09] Klipp, E. et al., *Systems Biology: A Textbook*, 1st Edition, Wiley-VCH, 2009.

[Koss11] Kossiakoff, A. et al., *Systems Engineering Principles and Practice*, 2nd Edition, Wiley-Interscience, 2011.

[Lasz96] Laszlo, E., *The Systems View of the World: A Holistic Vision for Our Time*, 2nd Edition, Hampton Pr, 1996.

[Luhm12] Luhmann, N., *Introduction to Systems Theory*, 1st Edition, Polity, 2012.

[Mann74] Manna, Z., *Mathematical Theory of Computation*, McGraw-Hill, 1974.

[Maie09] Maier, M. W., *The Art of Systems Architecting*, 3rd Edition, CRC Press, 2009.

[Mcke12] Mckeown, M., *The Strategy Book: How To Think and Act Strategically to Deliver Outstanding Results*, 1st Edition, FT Press, 2012.

[Mead08] Meadows, D. H., *Thinking in Systems: A Primer*, Chelsea Green Publishing, 2008.

[Miln89] Milner, R., *Communication and Concurrency*, Prentice-Hall, 1989.

[Miln99] Milner, R., *Communicating and Mobile Systems: the π-Calculus*, 1st Edition, Cambridge University Press, 1999.

[Mull11] Muller, G., *Systems Architecting: A Business Perspective*, CRC Press, 2011.

[Odum94] Odum, H. T., *Ecological and General Systems: An Introduction to Systems Ecology*, Rev Sub Edition, University Press of Colorado, 1994.

[Ogat03] Ogata, K., *System Dynamics*, 4th Edition, Prentice Hall, 2003.

[O'Rou03] O'Rourke, C. et al, *Enterprise Architecture Using the Zachman Framework*, 1st Edition, Course Technology, 2003.

[Palm09] Palm, W. III, *System Dynamics*, 2nd Edition, McGraw-Hill Science/Engineering/Math, 2009.

[Pele00] Peleg, M. et al., "The Model Multiplicity Problem: Experimenting with Real-Time Specification Methods". *IEEE Tran. on Software Engineering*. 26 (8), pp. 742–759, 2000.

[Pork78] Porkert, M., *Theoretical Foundations of Chinese Medicine: Systems of Correspondence*, The MIT Press, 1978.

[Pres09] Pressman, R. S., *Software Engineering: A Practitioner's Approach*, 7th Edition, McGraw-Hill, 2009.

[Raff11] Raff, H. et al., *Medical Physiology: A Systems Approach*, 1st Edition, McGraw-Hill Professional, 2011.

[Rayn09] Raynard, B., *TOGAF The Open Group Architecture Framework 100 Success Secrets*, Emereo Pty Ltd, 2009.

[Roza11] Rozanski, N. et al., *Software Systems Architecture: Working With Stakeholders Using Viewpoints and Perspectives*, 2nd Edition, Addison-Wesley Professional, 2011.

[Rumb91] Rumbaugh, J. et al., *Object-Oriented Modeling and Design*, Prentice-Hall, 1991.

[Sang03] Sangiorgi, D. et al., *The Pi-Calculus: A Theory of Mobile Processes*, Cambridge University Press, 2003.

[Scho10] Scholl, C., *Functional Decomposition with Applications to FPGA Synthesis*, Springer, 2010.

[Scot67] Scott, D. S., "A Proof of the Independence of the Continuum Hypothesis," *Mathematical Systems Theory*, Volume 1, 1967, pp. 89–111.

[Shap00] Shapiro. S., *Foundations without Foundationalism: A Case for Second-order Logic*, Oxford University Press, 2000.

[Shel11] Shelly, G. B., et al., *Systems Analysis and Design*, 9th Edition, Course Technology, 2011.

[Sher09] Sherwood, L., *Human Physiology: From Cells to Systems*, 7th Edition, Brooks Cole, 2009.

[Sode03] Soderborg, N.R. et al., "OPM-based Definitions and Operational Templates," *Communications of the ACM* 46(10), pp. 67-72, 2003.

[Somm06] Sommerville, I., *Software Engineering*, 8th Edition, Addison-Wesley, 2006.

[Voit12] Voit, E., *A First Course in Systems Biology*, 1st Edition, Garland Science,

2012.

[Warf06] Warfield, J. N., *An Introduction to Systems Science*, World Scientific Publishing Company, 2006.

[Weil00] Weil, A., *Spontaneous Healing: How to Discover and Embrace Your Body's Natural Ability to Maintain and Heal Itself*, Ballantine Books, 2000.

[Weil04] Weil, A., *Health and Healing: The Philosophy of Integrative Medicine and Optimum Health*, Revised Edition, Mariner Books, 2004.

INDEX

called port, 80, 86

calling port, 79, 84, 85

process algebra, 46, 73

   algebra of communicating processes, 74

   calculus of communicating systems, 74

   communicating sequential processes, 74

   infinite-queue SBC process algebra, 74

   multi-queue SBC process algebra, 74

   single-queue SBC process algebra, 74

process operation

   null process, 90

   parallel composition, 89

   recursion, 90

   renaming, 90

R

real system. *See* physical system

recursion, 90

rename, 135

renaming, 90

renaming combinatory, 90

renaming function, 90

S

SBC. *See* structure-behavior coalescence

SBC architecture, 51

   multi-queue SBC process algebra, 60

   SBC architecture development method, 58, 233

   SBC view model, 57, 231

SBC architecture development method, 58, 233

SBC view model, 57

   evolution&motivation view, 56, 58, 61, 64, 75, 141

   multi-level view, 56, 61, 75

   systemic view, 56, 61, 75

SBC-ADM. *See* SBC architecture development model

SBC-VM. *See* SBC view model

SC. *See* structure chart

second-order logic, 205

sequentialization, 89

single model. *See* model singularity

single-queue SBC process algebra, 74

SM. *See* systems model

SSA&D. *See* structured systems analysis and design

S-SBC-PA. *See* single-queue SBC process algebra

stakeholder, 37, 39

structural composition, 90

structural decomposition, 16, 66

structure chart, 52

structure-behavior coalescence, 50, 53

structured systems analysis and design, 52

   data flow diagram, 52

   structure chart, 52

system, 15, 19

   system dynamics, 15, 207

   systems analysis and design, 15

   systems architecting, 15

   systems architecture, 15

   systems bible, 15

   systems biology, 15

   systems ecology, 15

   systems engineering, 15

   systems medicine, 15

   systems modeling, 15

   systems physiology, 15

   systems requirement, 15

   systems science, 15

   systems theory, 15

   systems thinking, 15

   systems view, 15

systemic view, 31, 44, 46, 61, 75

   behavior view, 31

   input/output data view, 31

   interaction flow diagram, 58

   structure view, 31

252

systems architect, 41

systems architecture, 37, 38, 41

    definition, 37, 38

systems behavior, 32

systems definition, 15

    systems definition 1.0, 16

    systems definition 2.0, 63

systems model, 33, 34, 35, 36

    MVC architecture, 49

    SBC architecture, 54

    structured systems analysis and design, 52

    systems architecture, 37, 38, 55

systems structure, 32

T

target architecture, 42

transitional semantics, 101, 131, 223

V

view

    analysis view, 31

    behavior view, 31

    concept view, 31

    design view, 31

    evolution&motivation view, 31

    implementation view, 31

    input/output data view, 31

    multi-level view, 31

    strategy/version n view, 31

    strategy/version n+1 view, 31

    structure view, 31

    systemic view, 31

view model, 42, 44

virtual system, 19, 33

VM. *See* view model

www.ingramcontent.com/pod-product-compliance
Lightning Source LLC
Chambersburg PA
CBHW080652190526
45169CB00006B/2081